D1372836

Successful Product Development

Successful Product Development

Speeding from Opportunity to Profit

Milton D. Rosenau, Jr.
Rosenau Consulting Company
Bellaire (Houston), TX

JOHN WILEY & SONS, INC.
New York, Chichester, Weinheim, Brisbane, Singapore, Toronto

Library of Congress Cataloging-in-Publication Data:

Rosenau, Milton D., 1931-
    Successful product development : speeding from concept to
profit / by Milton D. Rosenau.
        p. cm.
    Includes index.
    ISBN 0-471-31532-X (alk. paper)
    1. New products.   I. Title.
    HP5415.153.R673   1999
    658.5'75--dc21

                                                                    99-30972

Printed in the United States of America.

10 9 8 7 6 5 4 3 2 1

Contents

APPENDICES

Preface

WHY ANOTHER BOOK ON NEW PRODUCT DEVELOPMENT?

A multitude of books on aspects of new product development have been written for practitioners and published in the 1990s. Many of the books have been concerned with accelerating time-to-market. Some books have dealt with particular functions (such as marketing), specialized aspects (such as market research), or separate tools (such as quality function deployment). Why another one? In short, this book aims to provide:

- A general overview of the entire new product development cycle
- Practical guidance on how to cash in more quickly on your firm's investment in new product development

This is not the same thing as getting to market faster, because that achievement—although obviously important—does not assure enough income to repay the development investment. This book is about how to reap profits quickly or, at least, soon enough to justify the investment. However, keep in mind that new product development is an idiosyncratic process. What works well in your company or your business unit may be unsuitable in another company or business unit. And any process you adopt can become bureaucratic, confining, and potentially, a bottleneck if it is not continually adjusted and carefully managed. Thus, you want to read this book to gain a perspective rather than a prescription. The lessons you choose to implement in your company will depend on your situation.

WHY IS NEW PRODUCT DEVELOPMENT SO CHALLENGING?

Despite the availability of much knowledge (and some firmly held opinions), more than one wag has commented that "many new product development projects are indistinguishable from hunting ducks at midnight without a moon . . . there's lots of shooting and squawking with only random results and a high probability of damage." This produces a tension in many companies. The tension between a business focus and a new product development focus arises because these two foci are fundamentally different.

The business focus, for which executives are normally compensated, is to produce existing products and services efficiently and reliably. This requires discipline, control, and predictability. A common focus for new product development is to produce a useful new product or service as quickly as possible. This frequently demands adaptability, flexibility, and the need to cope with the unexpected. Executives in companies that are consistently very successful at developing new products and services manage to balance these disparate emphases.

WHO THIS BOOK IS FOR

The intended audience is:

- All practitioners who are involved with any aspect of developing new products and services
- Managers of such practitioners
- Executives with responsibility for businesses that require new products and services

THIS BOOK'S APPROACH TO SPEEDING FROM IDEA TO PROFIT

I propose that there are five time-sequenced events and four intervals (or periods) between these events that must be mastered and understood to reap profits quickly from an investment in a new product effort. This is illustrated in Figure P-1. The central message of this book is that what is done—or not done—in the fuzzy front end (FFE) and stages and gates (S&G) intervals determines profitability after launch when product shipments start. In particular, it is advantageous for any company to shorten the length of time required to achieve their profit objective, that is, the duration of the preprofit sales

Five Events

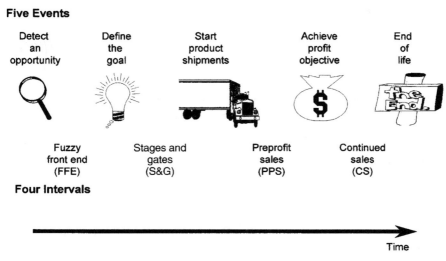

Detect an opportunity	Define the goal	Start product shipments	Achieve profit objective	End of life

Fuzzy front end (FFE)	Stages and gates (S&G)	Preprofit sales (PPS)	Continued sales (CS)

Four Intervals

Time

Figure P-1. *Five events and four intervals in the new product development cycle.*

(PPS) interval. (Options for defining the profit objective are discussed in more detail in Chapter 1.) Conversely, it is desirable to sustain or continue profitable sales as long as this is practical, to lengthen the continued sales (CS) interval. The first three intervals (FFE, S&G, and PPS) are potentially time consuming, and practical actions to shorten these intervals are described in the book.

USEFUL AND UNIQUE FEATURES OF THIS BOOK

My goal is to provide you with:

- A unified coverage of the new product development (NPD) process, from beginning to end
- A deliberately brief book, because busy practitioners have little time for a lengthy treatise
- Many figures to illustrate key ideas
- Summary points that are set at the page margins for emphasis and ease of recall
- A book that is based on the latest practical knowledge and research
- Numerous illustrative examples, mostly from current media

HOW THIS BOOK IS ORGANIZED

This book covers what you can do to shorten the overall time to profit and indicates some ways to cash in more quickly on your new product development investment. In it I discuss all four intervals, especially the FFE and S&G intervals. Although there may be some temporal overlap of the activities carried out in each of the four intervals, I separate their discussion in the book. There are six chapters, which are logically grouped into three parts:

Part 1: Introduction

1. Overview

Part 2: The New Product Development Process

2. The Fuzzy Front End Interval
3. The Stages and Gates Interval
4. After Launch: The Preprofit Sales and Continuing Sales Intervals

Part 3: Improving Your Process

5. Implementation
6. Continuous Improvement

In the overview we frame the issue, exploring what profit means and why a lack of clarity may make it elusive, and then define three intervals of time, from inception to profit. In Part II, Chapters 2 to 4, we explore the first three intervals and indicate key elements and actions that may be helpful in shortening time-to-profit. In Chapter 4 we also discuss actions that may be attractive after achieving the profit objective, that is, enjoying "gravy" and expanding on the product's profit.

Part III, Chapters 5 and 6, is about ways you can put some of the ideas into practice in your company. The fifth chapter covers implementing the lessons of the book. It deals heavily with resource allocation, which ultimately, is the key to accelerating your new product development payoff. Although resource allocation is also mentioned briefly in other chapters, the full discussion is deferred until Chapter 5. In Chapter 6 we explore the value of and ways to conduct continuous improvement reviews (often called *post mortems*), since these are the sine qua non upon which long-term improvement may be achieved. Supplemental information is provided in three appendixes: a list of abbreviations used in the book, the reference citations for each chapter, and a bibliography of other recent books on innovation and new product development.

Summary

The book details four important actions for your firm to consider:

1. Clarifying and agreeing on the profit goals that are your objective.
2. Managing the first three intervals (FFE, S&G, and PPS) to minimize the overall time-to-profit.
3 Allocating your necessarily limited resources to accelerate the highest-priority efforts.
4. Insisting on continuous improvement reviews so that future new product development efforts can be carried out better and faster.

MILTON D. ROSENAU, JR., CMC, FIMC

Rosenau Consulting Company
Bellaire (Houston), TX

Successful Product Development

Part 1

Introduction

1

Overview

This book is about cashing in more rapidly on your investment in new product and service development. It is written for practitioners who wish to improve the performance of their business by introducing a continuous stream of new products and services to the market. The book is concerned primarily with a sustainable product development process in a mature company. Startup companies developing a new, innovative entry to the market will also benefit.

In this chapter we discuss user obstacles that can delay and harm your intended profits, provide some basic background, and distinguish speed to profit from speeding to market. Following that, the five events and four intervals of the product development cycle are discussed, the profit objective is explored, and shortening time to profit is covered. This overview ends with a brief introduction to implementation.

An ongoing stream of profitable new products is better than a "shooting star," a single product or service that generates great publicity and excitement and then disappears.

OVERCOMING USER OBSTACLES

There are two extreme dangers in the management and execution of new product development efforts:

1. Making what you cannot sell
2. Selling what you cannot make

In the first, you are harmed by having aimed incorrectly, trying to push goods or services that are unsuitable or totally unsatisfactory for the intended user or market. In the second, the goods or services would be correct but you are unable to make them consistently, in adequate volume, or, because of technical difficulties or unavailable components, at all. My goal is to help you avoid both dangers.

The book is concerned about taking actions and making design choices *before* a new product or service is launched that make it easy—not hard, time consuming, or expensive—for customers to use the product or service *after* it is introduced for commercial sale. Early thinking about what happens after the product or service is sold and used can:

- Reduce warranty returns and expenses
- Decrease customer service costs
- Possibly obviate the need for a telephone help line
- Improve ease of use and generate favorable word of mouth recommendations
- Increase sales and profits

> **If your new product is hard to operate or maintain, it will discourage users and potential users.**

In short, what you do before launch to make your new product or service user friendly can both increase your profits (by reducing unnecessary costs) and shorten the time it takes to achieve your profit goal.

In early 1998, a colleague remarked that she was going to wait for several months after the then imminent launch of Windows 98 before buying a new computer with that operating system software installed. This computer-savvy woman is a graduate of both MIT and Harvard Business School and very much an early adopter of new electronic technology. Her reasoning was that the early commercial product, despite numerous beta test versions, would be flawed with troublesome bugs. Although she was better able to cope than most people, she saw no reason to become an unwilling product tester.

> **Errorware is worse than vaporware because it reaches the market and causes trouble for users.**

Just a few years ago, we used to joke about *vaporware,* that is, the announcement of new software that was, in fact, still being developed. In many cases it never even materialized at all. Now, we seem to have progressed to *errorware,* that is, software that is released but defective. Given the regrettable prevalence of errorware, only masochistic technocrats or gullible optimists are likely to be pioneering purchasers. Reluctance to purchase—especially if it is widespread—impedes product sales.

One company shipped new software and posted a note on its Web page listing the bugs they knew were in the product at the time of shipment [1].

As a result of suspected product defects, some potential purchasers delay adoption of many new products and services. (For simplicity, I will just refer to products in what follows, unless there is a significant difference where new services are concerned.) It's not that the new product (software in the case just cited above) would not save a user time or money (or provide other benefits of value), as illustrated in Figure 1-1. Rather, it is a realization that adoption will impose training and learning costs before the hoped-for benefits are obtained, as illustrated in Figure 1-2. For many people, the size of the adoption barrier is perceived to be so great that adoption delay, sometimes for years, is attractive.

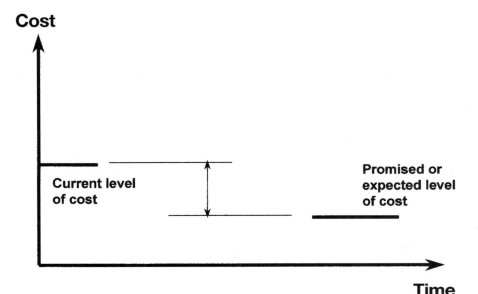

Figure 1-1. *Prospective savings on a user's adoption of a new product create an inducement to buy.*

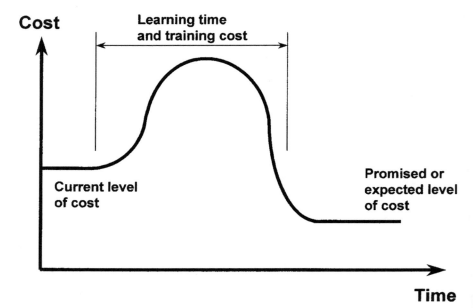

Cost

**Learning time
and training cost**

**Current level
of cost**

**Promised or
expected level
of cost**

Time

Figure 1-2. Adoption barrier due to user's expectation of costs that will be incurred before potential savings can be realized.

Returns of consumer electronics products exceeded 5 percent in 1995 and are believed to be increasing. In many cases, consumers were unable to figure out how to use the just-purchased products, and many of the retail stores from which they made their purchase were unable to help [2].

Buyers and users are balancing factors of performance benefits and imposed costs in deciding whether to commit to a new product. The size of the adoption barrier depends on a complex—and partially subjective—assessment of many factors. Some of the considerations about benefits are:

- The features offered
- The advantage promised by these features
- Availability of the product and any ancillary or adjunct consumables
- The prospective ease of use
- Promised or expected reliability
- Durability
- Convenience of service when required

In addition to the above, which apply to all prospective purchasers, the costs for business-to-business products include not only the monetary ones, but also the actions that will absorb management time in companies:

- Time required to get a purchase approval if the product is expensive
- Price negotiations
- Delivery arrangements and charges
- Setup costs, including any required facility modifications
- Training costs and time
- Indirect operating expenses (e.g., heat, air conditioning, electricity)
- The ongoing cost of consumables, spare parts, and periodic service

SOME BASIC BACKGROUND

Two questions deserve some discussion to provide a framework for what follows:

1. Why develop new products?
2. Why develop new products faster?

Why Develop New Products and Services?

I have heard many different answers when I asked participants in executive discussions or management workshops this question. Some of these answers are:

- The boss told me to do it.
- It's a corporate mandate.
- It's more fun than anything else.
- Someone came up with an interesting idea.
- It will help us maintain a proprietary position.
- It will smooth out highly seasonal product sales.
- We will be able to make better use of resources and production capacity.
- We have to match the competition.
- We have to respond to changed market conditions.

- It helps us to avoid obsolescence.
- It will permit us to expand the firm.
- We need to diversify.
- New products can increase the firm's profits.

Many of these reasons are obvious with just a moment's thought. However, some cases may reinforce the necessity to develop new products.

As an example of a corporate mandate, 3M now is obtaining 30 percent of its sales from products introduced within the last four years [3]. Their previous target was 25 percent of sales from products introduced in the last five years. Johnson & Johnson reported that 36 percent of 1997 sales came from products introduced in the past five years [4]. Hewlett-Packard has often reported over 60 percent of its orders were for products introduced in the last two years [5]. Although there is always ambiguity about the definition of a new product (e.g., is a minor product modification a new product?), these are three well-regarded companies that clearly demand and achieve a continuing flow of new products.

As another example, Figure 1-3 shows that obsolescence is inevitable. Consider the slide rule, which was previously ubiquitous for engineers and scientists. A top-of-the-line slide rule provided three significant figures for calculations, came with a sturdy leather case, sold for approximately $190 (in 1999 dollars), and did not require a battery. Today, we have portable scientific calculators that provide six to eight significant figures, cost about $8 (or perhaps even less), and do not require any cursor adjustment. Despite the need for a battery (minimized admittedly by inclusion of photovoltaic cells), slide rules have become obsolete and pocket calculators are in widespread use.

Setting of type—as exemplified by metal type manually set by craftsmen for the production of newspapers—has been made obsolete in that industry by the use of computers and computer printers. All the equipment required to support this ancient trade is now scarcely used any more [6].

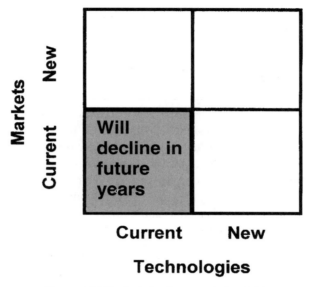

Figure 1-3. *Product obsolescence is inevitable.*

> Frank Popoff, chairman of the board of Dow Chemical Company, said: "The secret to success is a continuous flow of new products, processes and services that bring value to our customers. . . . America's best and most admired companies recognize the value of innovation" [7].

Why Develop New Products Faster?

Reduction of time-to-market is a major goal of many companies, and many firms have shortened development times very substantially [8]. There are at least a dozen reasons that firms have cited to answer this question:

- The firm can incorporate more timely technology than can slower firms.
- The firm can base the design on market research that is more current.
- The goal is more likely to stand still if launch is imminent.
- There is greater credibility with top management and in capital markets.
- There are lower development and investment costs.
- The firm can beat competition down the experience curve.
- Competition can be preempted.
- Customers may be locked in by imposing switching costs.

- More new products per year can be introduced.
- More sales are likely.
- More profits are likely.
- The firm is less likely to lose key people.

<table>
<tr>
<td>

There are many advantages to faster new product development.

</td>
<td>

In December 1997, Nissan announced plans to spend over $300 million to improve its computer-aided design capabilities. A major goal of this investment is to shorten the time to launch new vehicles to one year in 2002 from a bit less than two years at the time of the announcement. As reported, they hoped to achieve this ambitious goal by eliminating various prototypes and replacing or reducing crash tests by using computer simulations [9].

</td>
</tr>
</table>

WHY IS SPEEDING TO PROFIT AN ISSUE?

The widely practiced stage–gate new product development process is typically portrayed to end with "launch" and then there is a "pot of gold" or a "bag of cash," as illustrated in Figure 1-4. The concept for this process originated in the National Aeronautics and Space Administration many years ago, but has been effectively popularized by one writer (who has even trademarked the term *Stage–Gate*) [10]. In fact, if you have navigated successfully through the "stages and gates" obstacle course, there is nothing

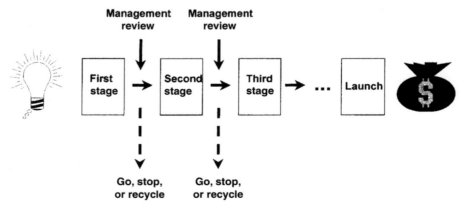

Figure 1-4. A general stage-gate process for new product development.

automatic about (1) making enough money to justify the risky new product development effort or (2) making money quickly enough to yield a satisfactory discounted cash flow return.

> One knowledgeable observer has commented: "Furthermore, many managers believe that they have a good process if they have been ISO 9001 certified. . . . [but] merely having a formal documented new product development process has no impact at all on performance. . . . The important thing is how the process is designed" [11].

But by focusing on the new product's launch (i.e., the time-to-market), companies develop new product myopia, overlooking the possible disappointment of lower sales than anticipated. In some cases there is a greater interest in getting the new product to market quickly than in delivering trouble-free functional utility and value to the intended user. Many developers fail to stress, or neglect, what follows launch. The technical support or customer service functions may be unable to answer users' questions, auxiliary equipment may be unavailable or in short supply, and although it is uncommon, consumable supplies required to operate the product may be hard to obtain. In many instances these kinds of avoidable deficiencies cause problems for innumerable users. And these problems impede the sales growth for the new product and may create a cost for warranty and service. Any user impediment postpones the time at which the company may reach its profit goal.

Quickly achieving the profit objective requires that the new product be easy to use by users, not just by the developers.

THE BIG PICTURE: FOUR INTERVALS AND FIVE EVENTS

Let's put the overall new product development cycle in perspective. There are four *intervals* or periods between detecting an opportunity for a new product and the end of a product's life, as shown in Figure 1-5:

1. The fuzzy front end (FFE)
2. The new product development (NPD) effort itself, which is typically conducted under a stages and gate process (S&G)
3. A period of initial, preprofit sales (PPS), ending with achievement of the profit objective(s)

Time-to-Market Is Often Critical
In many businesses, the first entrant dominates the market

Five events

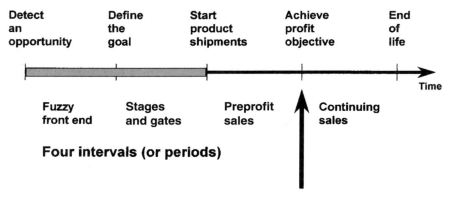

But Time-to-Profit Pays the Bills

Figure 1-5. Time-to-profit is longer than the time-to-market.

4. A period of continuing sales (CS) after the profit objective, which may be viewed as "gravy"

Many outstanding companies strive to reduce time-to-market. That is not to say that they have ignored improving time-to-profit, but it has not been the main emphasis.

In February 1998, Boeing was reportedly planning to reduce dramatically, to one year, the normal five-year time to develop a new airplane. Clearly, Boeing does not intend that the resulting new aircraft will have life-threatening defects. But the focus is on shortening the time to product launch, presumably driven by competitive market pressures [12].

Time-to-market may be the right goal because of competitive and market considerations. In some markets the first company to have salable products can later dominate the market.

> The market for contact lenses and the concomitant sterilization fluids provides one example in which time-to-market is paramount. This is a fiercely competitive market. The companies that develop and sell these products are convinced that the company that is first to market will have the dominant market share. Users are reluctant to switch within a product class (such as flexible long-wear lenses). Users are inclined to switch only when there is a totally new product class (such as extended-wear lenses) that promises them new benefits not obtained with their use of the present product class.

When development time is stressed, some companies fail to distinguish the specific objective and the rationale or conditions that justify this emphasis. One study [13] examines the conditions under which the development goal should be:

- First to market
- Fast follower
- Follower

The best choice turns out to depend on several factors:

- The company's competitive position
- Market size
- Market growth rate
- The expected performance level of the product
- Development cost

Clearly, faster development speed is not automatically the most desirable choice in every situation.

In any event, shortening time-to-market may be the wrong race unless the developer can also win the time-to-profit race. Until a firm can consistently achieve its new product profit objective, it runs the risk that it will eventually fail. This is a pitfall encountered by some shooting stars. There is no advantage in speed for speed's sake. The wrong product fast or the right product slow are both less desirable than the right product right with the shortest possible time-to-profit.

Your goal should be to shorten time-to-profit, not merely one or two of the intervals before the profit objective is achieved.

The situation is like a triathlon. The winner is not the person who is fastest in one or two portions of the race. The winner is the one who has the best overall race time. That winning person may well be second, third, or farther back in the pack in each of the three event sections. The same is true of new product development. The one who wins the race to profit may not be the fastest with the fuzzy front end, the stages and gates, or the preprofit sales intervals. It is the combined result of all three of these that is central to winning the race from the start (detecting a new product opportunity) to the finish (achieving the profit objective you have set for it).

THE PROFIT OBJECTIVE

Researchers have discovered that the new product profit objective is normally *multidimensional,* having both *project-level* and *firm-level* dimensions or components [14]. Their findings are summarized in Figure 1-6. There are often two simultaneous project-level measures, which may be comprised of a combination of market share, revenue, profit, competitive advantage, or customer satisfaction or acceptance. The firm-level objectives can include both a financial and a performance measure, such as return on investment, percent of profits from new products in recent years, the success versus failure ratio, or a fit to business strategy.

A firm's profit objective may be multidimensional.

Thus, for a firm, their new product profit objective may be stated as including:

· A stated amount of profit dollars
· A particular sales level

Project-level
  - **Market share**
  - **Revenue**
  - **Profit**
  - **Competitive advantage**
  - **Customer:**
    • **Satisfaction**
    • **Acceptance**

Firm-level
  - **ROI**
  - **Percent of profits from new products**
  - **Success-failure rate**
  - **Fit to business strategy**

Figure 1-6. Dimensions and measure of the new product profit objective.

- A specific return on investment
- Some explicit market share goal, often to be the leader [15]
- Some other proxy for profit that is appropriate to the situation

Many companies have a financial goal for a new product development project, which may be stated as a net present value (NPV) or internal rate of return (IRR). NPV and IRR require the selection of a time horizon for the discounted cash flow analysis that is required for their calculation. That time horizon could be the desired (or mandated) time-to-profit. Other companies use the break-even time. The problem with break-even time is that getting only to break even does not compensate a company for the risk it has assumed in undertaking the development. To put this differently, all that break-even time tells you is when you can expect to get back to zero investment, not when you can expect to reach a desired level of profit. The proper objective of new product development is to earn an appropriately large return on the inherently risky effort and to do so quickly enough to promise an attractive investment.

Hewlett-Packard has publicized its use of a Return Map. This semilogarithmic graphic device plots investment and return (on the vertical logarithmic axis) versus time (on the horizontal linear axis). Although they supposedly find it difficult to use this tool in every situation, their immediate goal has been to shorten break-even time and then achieve a fourfold return on the total investment [16].

Different companies have different attitudes about risk. Regulated businesses (e.g., regional telephone companies, utilities, etc.) tend to be conservative and risk averse. Other companies, such as those in fast-moving fields, are often more aggressive and tolerant of risk. As a general rule, you should aim to earn a higher level of profits than your normal target (however this is measured) when the new product development effort has greater than normal risk.

To stay in the lead in its fiercely competitive high-technology market, a company like Intel must commit more than $1 billion to build a production facility for a new chip that is not fully designed.

The profit goal may apply only to a specific new product or service, or it may apply to the synergistic impact such an effort has for the company. The goal of an effective *overall* product development effort should be the rapid achievement of the profit goal. This may—or may not—mean that the FFE and S&G intervals are accelerated.

SHORTENING TIME-TO-PROFIT

Recently, it has become increasingly apparent that there are additional ways to improve *time-to-profit,* which was and still is the driving reason to shorten time-to-market. Figure 1-7 shows that there are three ways to shorten the overall process. You can shorten the time-to-profit by shortening all three intervals (e.g., uniform reduction like a rubber band). Alternatively, it may subsequently be possible to lengthen the FFE interval to shorten either or both of the S&G and PPS intervals. Finally, it may be possible to lengthen the S&G interval to shorten the PPS interval by a greater amount. In the following sections we discuss each interval.

Shortening FFE (Chapter 2)

Shortening time-to-market (i.e., the time of launch at the end of the S&G interval) is helpful. One way to do this is to seize opportunities to shorten the fuzzy front end. Some of the ways to do this:

- Know and apply the company's strategic screens to both the idea search and to any identified ideas.

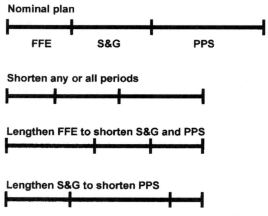

Figure 1-7. Three options to shorten time-to-profit.

- Execute the screening rapidly rather than permitting marginal ideas to float around, consuming the limited time of your precious human resources and diverting attention from management's communication of the high priorities.
- Establish a storage repository for attractive ideas that are not timely (e.g., because resources are not available to examine them).

All of these time-savers (which are covered more thoroughly in Chapter 2) require that the company have appropriate resources available for and dedicated to the FFE interval. To be effective, the people must represent all critical functions, and the available facilities must be appropriate for the nature of the business.

Shortening S&G (Chapter 3)

For several years, many of us have advocated that companies must shorten the time-to-market for developing new products. The numerous benefits of a speed improvement (listed above) have been well documented. For many people, shortening time-to-market was synonymous with speeding up the S&G interval. Until recently, although the opportunity was identified, shortening the FFE interval was not given much systematic attention.

> *Adequate resources are required to shorten any interval.*

Obviously, acceleration of product launch (i.e., shortening time-to-market) never meant sacrificing quality to get to market quickly, nor did it imply that critical development steps should be omitted. Rather, the emphasis was—and should remain—on completing the S&G interval as quickly as possible to gain the many benefits of being first to market (or nearly so).

Regrettably, many firms have installed a S&G new product development process, but the new product development teams lack a clear understanding of either the firm's new product development strategy or the multidimensional profit objective. In too many instances, the S&G interval starts but the FFE work is not complete, so important elements (e.g., some of the requirements for the new product) are unsettled. Lacking knowledge of the desired destination, a new product development team is dependent on luck to be successful.

Shortening PPS and CS (Chapter 4)

Similarly, there are ways to shorten the duration of preprofit sales required to achieve the profit objective. Most of these depend on advance planning of the product launch, including ways to achieve rapid distribution in high-margin channels. Obviously, the choice of product features and price point are key elements. Similarly, the way the sales force is introduced to the product and the opportunity it provides them to excite customers can be very influential. Perhaps most important is the product's ease of use, since favorable word of mouth can do wonders to generate demand.

The CS interval can be an anomaly. In many cases you want to *lengthen* the CS interval, assuming that it remains profitable. In some cases, regardless of profitability, you want to lengthen it to retain satisfied customers, either for goodwill or to hold them until a new product is ready.

IMPLEMENTATION (CHAPTER 5) AND CONTINUOUS IMPROVEMENT REVIEWS (CHAPTER 6)

A new product development schedule is based and depends on adequate resources, which must be provided by executive management.

Chapter 5 is devoted to steps you can take to shorten time-to-profit. The key to this is how a firm's finite resources are allocated. A portfolio management system is intended to assure that limited resources are applied to those efforts that make the most sense. Some elements of a portfolio management system include:

- A standard attractiveness rating scale:
  - For varied product development efforts that deal adequately with individual projects that are currently in different intervals
  - With uniform standards for judging which efforts to initiate and which to continue when they are at different states in the overall process
  - That establishes the priority of efforts
  - A culture that assures a way to stop work on low-ranked efforts when resources (mostly human, but sometimes physical) are insufficient
- Some plan for allocation across investment opportunities, as outlined schematically in Figure 1-8, such as:
  - 60 percent for simple development efforts (e.g., product modifications, me-too enhancements, or line extensions)
  - 30 percent for new platforms that will be the basis on which new products are based in the coming years
  - 10 percent for new-to-the-world or radical breakthrough efforts

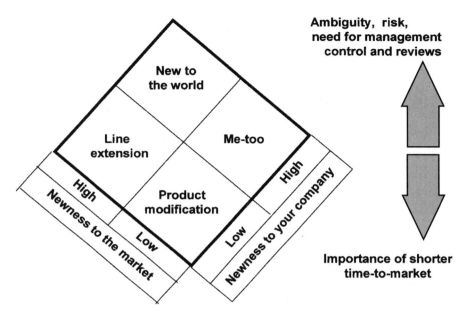

Figure 1-8. A portfolio of new product investment options must be distributed, and different kinds of projects require different kinds of management attention.

The criticality of faster time-to-market is altered depending on which quadrant your NPD effort is in. In general, speed is more important for imitative products and less so for truly novel products (unless other events are reducing the demand and future opportunity). There are two simplifying assumptions in Figure 1-8. First, a resource allocation scheme exists that must assure adequate resources for projects where speed is critical, and this scheme—once established—may not require excessive executive time. Second, since any new-to-the-world development frequently requires iterative work and planning (do a little, see how it works, plan the next steps), it normally requires frequent executive attention to adjust resource availability to match new needs.

The bulk of the examples and cases that are cited in the book are for discrete manufactured products. Therefore, in Chapter 5, I also discuss how other new product development situations differ from discrete manufactured products. This will show that the general lessons are also applicable to nonassembled goods, software, radical product innovation, services, and other situations.

Chapter 6 is devoted to a discussion of continuous improvement reviews. Whether these are called postmortems, postcommercialization reviews (or audits), or something similar, they are the most frequently omitted part of an effective new product development process. It is difficult, if not impossible, to improve unless resources and time are devoted to an objective examination of both completed (and in-process) development projects and the process itself.

SUMMARY OF KEY POINTS

- Obstacles to ease of use can delay achieving your time-to-profit objective.
- Faster new product development is generally beneficial but is not the same thing as faster time-to-profit.
- Faster time-to-profit is a—perhaps the—key objective you should pursue.
- The profit objective is multidimensional with project-level and firm-level components.
- There are five events and four intervals that must be managed.
- There are three different ways to shorten time-to-profit.
- You must implement new practices to achieve shorter time-to-profit.

Part 2

The New Product Development Process

2

The Fuzzy Front
End Interval

INTRODUCTION

The fuzzy front end (FFE) interval is the first element of a defined, fast process:

- It starts when someone detects an opportunity.
- It ends successfully when:
  - Enough is known about the opportunity so that the goal for the S&G interval is clear, including:
    - The product's requirements and proposed performance
    - The launch schedule
    - The ultimate profit objective and its timing
- The concept does not contain major obvious unknowns about the market, technology, or production process.
- A plausible business case for the product can be made.

To put this differently, a key FFE objective is to set the S&G interval's goal before an ill-considered or marginal idea is developed into finished form or launched into the market with great energy and at substantial expense. The FFE interval may also end earlier

The team that is working on a FFE effort must be big enough and have enough support to work through the issues that must be resolved.

if it becomes clear that the detected opportunity is not, in fact, sufficiently promising and it is dropped from further consideration. For the FFE interval to be successful, adequate resources (human and physical) and a realistic schedule must be dedicated to the effort, and it therefore must command an adequate budget.

It is easy to detect or articulate opportunities that are superficially attractive. Sorting out the wheat from the chaff is more difficult. Very few new product or service ideas are impossible, that is, clearly precluded by:

- The laws of physics (e.g., the limitations imposed by the speed of light or the properties of currently known materials)
- Regulation (e.g., antitrust)
- Your company's inherent limitations (e.g., access to capital funds)

The FFE is the "fish or cut bait" stage.

Thus, the challenge is to pick worthwhile and achievable targets. That's what the effort in the FFE is all about.

> Baxter Healthcare has a trademarked RIGHT–products–RIGHT slogan. The FFE interval is concerned with gaining an assurance that you are doing the right products, that is, picking suitable targets. The S&G interval, in distinction, is concerned with doing the product right, that is, executing it so that its development contains a suitable combination of user benefits.

The FFE is not a unidirectional series of steps as typically encountered in any stage–gate process, such as that illustrated in Figure 1-4. The things that have to be done and the sequence in which they must be done depend on the kind of opportunity stimulus (see below) that led to initiating the FFE work. Although the end objective is clear—removal of unknowns and development of a promising business case—the way this is achieved can and often does involve a lot of "backing and filling" as one route or aspect after another is investigated and trade-off choices are made.

The FFE interval may be disorderly, but it consists of goal-directed activities.

It is also important to realize that the FFE interval is not the same thing as capitalizing on technology change or merely conducting interesting (even exciting) scientific research. Although an opportunity may be detected as a result of systematic research or an accidental discovery, the opportunity may arise from three other sources.

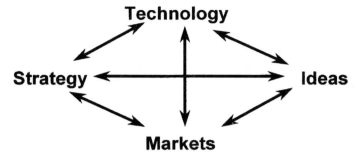

Figure 2-1. *The four critical FFE factors.*

FOUR FFE FACTORS

Figure 2-1 shows clearly that you can initiate the FFE as a result of any of four stimuli. But all four of these factors (strategy, ideas, technology, and markets) must be considered carefully and satisfactorily resolved in the initial FFE interval. The objective and intent of the formative period is to remove uncertainties, because a new product development schedule is not credible if there are major unknowns about the market, the technology, or the production process. Failure to execute the FFE carefully is often the cause of extra, more costly effort during the subsequent S&G or PPS intervals.

The model shown in Figure 2-1 is disarmingly simple while being general and therefore potentially very useful [1]. For example, either market or technology discontinuities can provide excellent entry points into initiating the FFE. Senior executives may want to change the firm's strategy (but only thoughtfully and carefully) as a result of a new, "killer" idea. The "window of opportunity" for any effort that starts the FFE must be considered early and may lead to an early termination or require the acceleration of an effort. In all cases, the end of the FFE is a solid business case that has been judged against preexisting and understood criteria.

As one cynical observer said, "an extra year of development will save you one month of homework."

In the late 1970s and early 1980s, many of the major oil companies invested in the development of various photovoltaic devices to convert sunlight to electricity. These development efforts were initiated as a hedge against the possibility of diminished access to oil. Most of these

(continued)

photovoltaic efforts languished when oil supplies were again readily available, because the newer photovoltaic energy sources were neither sufficiently efficient nor economically attractive to justify continued product development efforts.

In the earliest portions of the FFE interval, deciding which opportunities to push and pursue and which to deemphasize is not an objective decision. At such an early time, not enough is known with certainty to be assured that a selection will be optimum much less proven adequately profitable. Nevertheless, choices must be made and resources must be committed, so senior executives must actively participate in the screening process.

The FFE interval requires a high tolerance for ambiguity.

The varied things that must be resolved in the FFE interval demand that all functions must be involved. The FFE is not solely the province of marketing or the technologists (research, engineering, software, chemistry, etc.). These functions commonly are heavily involved, but so are manufacturing, production, procurement, sales, quality, customer service, and finance. It must be a multifunctional activity.

To shorten the FFE interval it is necessary to assign as soon as possible at least a small multifunctional team to investigate any superficially attractive opportunity. This team must then quickly examine the opportunity to decide if it is sufficiently attractive to justify further, more extensive exploration, as illustrated in Figure 2-2. Depending on the details of the opportunity, the team may require a day, a week, or possibly, a few weeks for this first examination. A promising opportunity may then be explored by a larger multifunctional team for a period of time that is a few times longer (say, three to five times longer) than the preliminary first examination. The notion is to commit increasingly longer, but strictly time limited periods to examining carefully opportunities that continue to have merit.

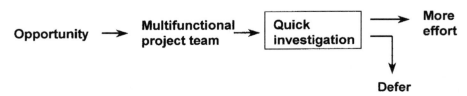

Figure 2-2. Opportunities must be quickly sorted in the FFE.

Fluke Corporation commits 100 days and $100,000 to fund the examination of designated new product opportunities, which includes producing an initial proposal for these. An additional 100 days can then be allotted to refining the business plan for the most attractive proposal(s). At that point, a promising effort can be fully authorized [2].

Some companies perform a discounted cash flow (DCF) analysis (net present value, NPV, or internal rate of return, IRR) during this early FFE interval. This is premature, because you don't know what you don't know at the earliest stage. Decision and risk analysis or critical assumption planning, like DCFs, are highly useful but should really be done only after some market assessment and development work has been done. Otherwise, a forceful advocate can justify almost any effort.

A peripheral equipment division of a major telecommunications company had to conform to the parent firm's new product development process. This process was created for the development and introduction of major telecommunications systems that required massive investments. As a result, the parent firm's process mandated a DCF analysis in the very early FFE effort. Regardless of the (questionable) merits of this requirement for the parent firm, it was premature to include such an analysis at an early point in the division's process.

Reliance on a "back of the envelope" calculation is more appropriate at very early stages. Do a quick iteration of the four FFE factors, for example, and then make a rough benefit-to-cost estimate.

In the mid-1970s, a researcher proposed a $100,000 development program to formulate a new compound for use in one aspect of car manufacturing. As development programs went in this company at that time, this was a fairly large effort. To conceal the specifics, pretend that the compound—if successfully developed—would provide cabling insulation, that the then selling price of comparable compounds was $3 per 1000 feet of cabling, and that each car then used about 35 feet of cabling. At the time, approximately 10 million cars were manufactured in the United States annually. Therefore, if the company could capture 100 percent of the market, annual

(continued)

---

sales would approximate $1,000,000. However, the company traditionally captured only 10 percent of similar markets, which would mean sales of about $100,000. Then, if the compound were as profitable as similar company products, it would yield a $10,000 before-tax profit. Thus, the company was faced with the certain expenditure of $100,000 and the uncertain prospect of earning $10,000. This simple "back of the envelope" estimate was sufficient to conclude that the proposed development was not attractive and should be abandoned or redirected.

In particular, the probable size of the opportunity and the cost of getting there is worth estimating. In the case of a formal financial justification during the FFE for a very innovative new product, it is virtually impossible even to make educated estimates because so much is unknown, multiple options are being examined, and all schedules are uncertain.

Very innovative efforts are characterized by uncertainty and you may not be able to demonstrate convincingly anything less than an infinite time-to-profit at an early stage.

Considering the heavy involvement of both marketing and technology personnel in the FFE interval, it is helpful to contrast "market pull" and "technology push" as ways to start. Both can be successful. If the effort arises from market pull, the main challenge is to create a product with suitable value. If the effort arises because of technology push, the challenge is to create and educate the market that the new technology is useful and appropriate. Figure 2-3 illustrates what is involved de-

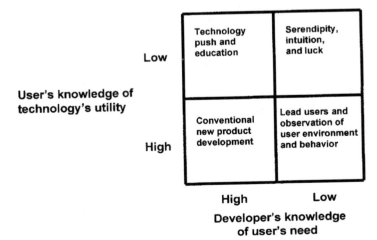

Figure 2-3. *Developer and user knowledge.*

pending on the developer's knowledge of the user's need and the user's knowledge the technology's utility.

Some companies (such as Hewlett-Packard, IBM, Intel, Motorola, and Xerox) have social scientists who live with a user for a day (or longer) in the user's environment to observe actual behavior. The goal is to learn what a user is really doing rather than rely on what a user reports that he or she is doing. Such observation may uncover ways that users overcome product limitations or compensate for inadequacies [3].

In the next four sections we discuss each of the four potential stimuli for a FFE effort. Keep in mind that any of these four may arise from an offensive (proactive) or defensive (reactive) stimulus. In the fifth section we discuss hybrid situations.

Technology Stimulus

The classic image of an opportunity arising from technology stimulation is something that comes about from an output of fundamental research, a serendipitous discovery, or a clever, original invention. But people do not purchase or use technology per se. They buy only the cost-effective benefits that technology may provide.

As one example of serendipity, Clarence Birdseye allegedly recognized the opportunity for quick-freezing fresh food after observing that the almost instantaneously frozen fish he caught while ice fishing tasted fresh after being thawed and cooked [4].

The first practical application of the red helium–neon laser was to assist in laying sewer pipe so that it was in a very straight line. This improved straightness assured a better flow and permitted the use of a narrower-diameter pipe.

New advanced technology may solve mundane problems.

In some cases, a considerable education effort may be required to explain to potential users why and how the new technology can provide a benefit. Even something as simple as a toothbrush—in this case the new $5 Gillette toothbrush—may require consumer education about why a price that is twice as high as for other toothbrushes is justified [5]. In fact, many new and interesting technologies do not reach the market at all, and some that do are not commercially significant.

You must expect to educate users about the benefits of any radical new technology.

Shape-memory metal alloys change from one shape to another shape (that is stored in its "memory") at a specific temperature. This intriguing technology has been known for a few decades and has been applied occasionally in the U.S. space program, including the first lunar landing. Other applications are occasionally proposed [6]. But the technology's availability has not yet led to major commercial applications.

Electromechanical film is being developed in Finland to produce very thin, flexible loudspeakers or microphones. Whether this exciting technology can succeed commercially is still unclear [7].

No matter how clever or interesting, technology does not itself create a market need. Examples of intriguing technologically driven new products that were commercial failures include the Newton MessagePad, Picture Phones, SelectaVision, and many others [8]. Bucky-balls are an example of very exciting technology that has so far proven too expensive to exploit for many promising applications [9]. Electronic "ink" that can change color [10] and plastic light-emitting diodes [11] are two new technologies for which commercially viable applications are presently unclear. Ultrawideband radio (also known as digital pulse wireless) is a new technology for which regulatory approval by the Federal Communications Commission is required before this can be considered seriously for varied interesting commercial opportunities [12]. If there is a latent need that a new technology can help satisfy, it may become the basis of a commercially important new product.

New technology alone is not a sufficient justification for a new product development effort.

Better image quality is an example of a long-standing latent need for improved precision lenses. Aspheric lenses (i.e., those that have a non-spherical but rotationally symmetrical surface) can provide a better quality image in a camera or telescope than is otherwise obtainable with lenses having purely spherical surfaces. Aspheric lenses used to be quite difficult to manufacture in glass but quite easy to manufacture in plastic (once a master mold was precisely made). High-quality cameras and telescopes require glass lenses to perform well over a large range of temperatures for which plastic is unsatisfactory. With technical advances in computer-controlled machinery, glass lenses with aspheric surfaces can be manufactured with relative ease, and cameras and telescopes can now provide better image quality.

In early 1998, Intel created its new AnswerExpress service to provide telephone help to solve subscriber's problems with any manufacturer's computer hardware or software. The idea for the service depends on an interrogation technology that permits Intel to gather pertinent information about a subscriber's hardware and software while their computer is connected to Intel's network [13].

Many companies with large research and development or engineering staffs are primarily technology driven. This large body of technical talent is constantly creating interesting possibilities. In many cases the technologists stimulate interest among colleagues, and a large amount of energy is invested in a random examination of the new possibility. This amounts to an undirected, unchanneled FFE interval effort.

As one example of a company that has developed a systematic approach to the FFE, Eastman Kodak previously started with an idea (generally from a technologist) and then went directly into the stages and gates development process. They now select an opportunity, then go through their "discovery and innovation" process (which entails forming an innovation charter, seeding ideation, building models, samples, and prototypes for screening, and developing a business plan), and only then initiate the stages and gates part of development. The old and new approaches are contrasted in Figure 2-4.

- **Old process**
  - **Idea**
  - **Stages and gates**

- **New process**
  - **Select opportunity**
  - **D&I process**
    - **Innovation charter**
    - **Seed ideation**
    - **Build and screen**
    - **Plan**
  - **Stages and gates**

Figure 2-4. *Kodak's discovery and innovation process.*

> *Only employ appropriate technology where it can solve real problems in a cost-effective way.*

Technology has other roles to play in the FFE interval. The technologists can provide options to respond to opportunities that are otherwise identified. Or, they can rapidly carry out experiments to help optimize the preliminary characteristics of a new product. Technologists may employ rapid prototyping or other modeling methods to facilitate essential design formulations.

Considering the importance of technology, the question arises whether it must be confined to that which is available solely within your company. In some situations, this can be a classic "make or buy" question. The short answer is that you should not be locked exclusively into purely internal technology. Some of the issues bearing on the use of external technology are illustrated in Figure 2-5.

Market Stimulus

The classic image of initiating the FFE interval from a market orientation is to detect an unsatisfied need and figure out how to fill that need before anyone else. Ideally, the detected opportunity will be one you can uniquely satisfy with proprietary knowledge, thus creating a barrier to entry of competition.

> Postage is now being offered for sale in test markets over the Internet by E-Stamp Corporation. The company believes that small businesses, which presently do not make much use of postage meters, represent a

(continued)

	Do own R&D	Team with a company that is already expert
Reasons for	Own all results Build expertise	Can be faster Less investment
Reasons against	Can be myopic NIH	Must know enough to buy intelligently Commitment to use other's technology

Figure 2-5. Pros and cons of a firm doing its own R&D versus teaming with another company.

huge market. Postage is a market regulated by the Postal Service to prevent theft or falsification of a monetary instrument. This regulation will limit to some extent the amount of competition that eventually emerges if or when the market for electronic stamps develops [14].

Theft of both beverages and money (and the waste of beverages) is evidently a prevalent problem in bars. Presumably this is a real problem, the solution of which has a calculable monetary benefit for a bar owner. One entrepreneur has devised an electronic monitoring device for each bottle that will feed data to a central computer, thus monitoring and recording for analysis the specific actual performance of each bartender. Although expensive, it seems to be cost-justified at high-volume bars [15].

Secure identification of people making financial transactions at, for instance, automated teller machines, is currently dependent on personal identification numbers. This is also the case for many Internet transactions. One company has developed technology to record and then use a person's fingerprints instead of a personal identification number [16].

There are, however, needs—sometimes substantial and very important—that may not represent attractive new product development targets. For example, prospective sales income may be limited, or distribution channels may be unattractive.

A drug to prevent or cure malaria illustrates a product for which there is a real market need. Whether those people who are most in need of the drug, typically residents of the rural tropical endemic areas, can afford the drug (if it could be developed) is very unclear. Alternatively, a drug developer—if successful—might try to sell the new drug through some international relief organization, but such a channel of distribution may also limit the potential profits.

It appears that office printing increases very substantially when e-mail is newly installed. Although interesting, this observation does not point to the clear need or opportunity for any obvious new product.

If you do not suddenly detect a latent, undiscovered need, for which you only have to market what will be an eagerly sought after product, you always have the possibility of creating a market. In the latent need case, the market is waiting for the new product (although it may not yet realize it), but in the case of creating a market need, you have to invest to educate people about the new capability.

Some unsatisfied needs do not lead to attractive market opportunities.

Regardless of whether you start from a technology or market stimulus, you will serve the market better if you plan on launching a family of new products. New product development maps can prove helpful for this [17]. The family approach is very common in the field of consumer electronics. There are several potential benefits:

· You can initially launch the simpler models that can be introduced rapidly.
· Subsequent members of the product family may require only a few new components, subassemblies, or production processes, which can reduce your development risk and be easier to introduce quickly.
· The family of products should be able to use many common components, subassemblies, and production processes.

- Parts commonality can reduce your required inventory.
- These early models will give you a real presence in the market, far more revealing than any market research, and user feedback will help you better optimize later models.
- The family may permit you to offer slightly different models that reach many more users in your traditional channel of distribution.
- The family may permit you to distribute slightly different models through distinctively different channels of distribution, thus serving diverse markets.
- It is easier to defer "feature creep" and its attendant delay to later models by targeting and concentrating on a simple but useful initial product.
- A successful family can provide a "hook" to extend your family's lifetime, since some users like—perhaps even depend upon—the notion that you will be in the market for a long time and will be able to supply their changing needs.

One marketing consultant counsels that you should work to anticipate changing customer needs and invest enough to upgrade your own products before they become obsolete [18].

Hewlett-Packard's HP DeskJet 820C printer is the first HP inkjet printer in an evolutionary product plan that takes advantage of computer and operating system trends to make inkjet printing affordable for more users. . . . We realized that no single product program could successfully satisfy all of these criteria, so we needed to develop a phased approach. We decided that each new product development should leverage previous capabilities while incorporating a small set of new and innovative capabilities focused on our customer needs. These new capabilities would then be leveraged forward into succeeding efforts. In this fashion we could ensure a timely series of product introductions, each building upon previous successes and incrementally providing new capabilities that would ultimately satisfy all of our strategic initiatives. In addition to the market timeliness gained by a phased approach, we also knew that this plan would use scarce resources in the most efficient manner [19].

Source: Hewlett-Packard Company © 1997. Reproduced with permission.

Product families are a tactic previously employed by Hewlett-Packard. This widely admired company works to obsolete its own products with better replacements before competitors can [20]. For example, when they first released the HP Pavilion computers in late 1995, they released six models simultaneously. These differed in providing varied combinations of processor (AMD 486, Intel Pentium 75 MHz, Intel Pentium 90 MHz, or Intel Pentium 100 MHz), hard drive (635 MB, 850 MB, or 1260 MB), standard RAM (8 MB or 16 MB, all expandable), primary cache (0 or 256K, all expandable), and so on for other features. As is now apparent, the company has continued to enhance capabilities far beyond those in these introductory family members.

Vivitar simultaneously released two models of 110 cameras in the mid-1970s. Except for the logo on the front and an extra part in the cheaper camera, these were identical. The extra part was a tiny pinhole stop in the Model 20's lens to reduce its best photographic performance from F/4 to F/11. The commonality of other parts dramatically reduced the factory cost of both models. The Model 20 had a list price of $19.95 and the Model 30 had a list price of $29.95.

Product families, a series of similar products built on a common platform that provide varied combinations of benefits, offer many advantages.

At the time of its introduction, the CrossPad appeared to be aimed at satisfying an unsatisfied market need. It did not seem to be part of a product family, although there may well be further development plans that are not yet apparent. This device allows a user to write on a conventional pad that automatically records the notes in a computer file, which can be viewed or edited later. The notion is to avoid the necessity to transcribe handwritten notes and obviate the distraction of using a computer while participating in a meeting. When introduced, only one model was offered and it remains unclear whether this clever idea will be further developed to overcome its limitations, satisfy multiple potential needs, and have an array of price points [21].

Perhaps the opportunity will be detected as the result of market research carried out as a conscious proactive search. Or the opportunity may be detected as an accidental by-product of market research that has been carried out for some other reason. A full treatment of market research techniques is beyond the scope of this book, but it is important to understand that these may be qualitative or quantitative. A focus group is an example of the former, and conjoint analysis is an example of the latter. New variations and improvements in all market research techniques continue to emerge [22].

> The interactive concept test is a way to provide respondents in a conjoint analysis research study with dynamically changing prices as they consider varied combinations of potential product features. This improvement over conventional conjoint analysis more faithfully reproduces the potential purchase environment [23].

Some market research techniques, of course, are applicable only to product concepts that are within the experience of the subjects. It is very hard to get meaningful market research results for any new-to-the-world product, since users may be unable to visualize or understand the concept [24].

In the case of radical innovations, the needs of users, appropriate distribution channels, and the nature of potential competition may be uncertain [25]. Prospective users, unfamiliar with the concept for a new product that is a radical innovation, may only focus on familiar aspects [26].

Market research is inevitably limited to three sorts of hard data:

1. What people have done in the past
2. What people do now
3. What people say now

What we want to discover, however, is what people will do in the future, when our new product is available. It is also important to recognize that interest in a new product concept, or even an expressed purchase intent, is not—and may never become—an actual purchase. It is therefore important to examine the entire new product development supply chain, the concept of which is illustrated in Figure 2-6. In this figure, customers are the buyers, whereas users are those who actually employ the product in some application. In the case of home use of personal computers, Dell Corporation has organized direct distribution and sales so that a user is generally the same person as the

Exploitable opportunities may arise when competitors are constrained by structural limitations such as an existing channel of distribution.

Figure 2-6. New product development (NPD) supply chain.

customer. That is, the retail channel intermediaries have been eliminated and Dell is able to exploit a comparatively advantageous distribution channel. Although there are other cases that are similar, in many cases there are channel intermediaries who are the producer's customer and who are not the ultimate user. There are some situations where the customer and user are inherently different, such as pet foods and baby products. For many business-to-business products, the customer is a group and the purchase decision is made by several people, sometimes without any participation users.

Figure 2-7 provides a specific illustration for the package printing industry [27]. The printer and converter should be talking to all the ink producer's suppliers to understand what might be emerging. Similarly, the ink producer should be talking to both the packager of goods and the various elements

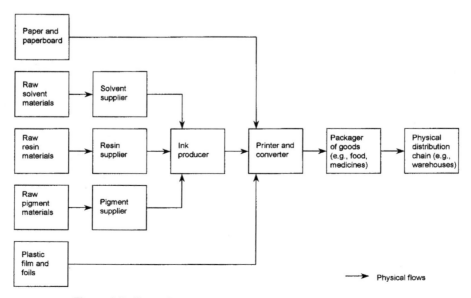

Figure 2-7. Illustrative value chain for the package printing industry.

of the physical distribution chain to better learn about the emerging needs of ultimate users and handlers.

As shown in Figure 1-5, recall that you may develop new products that serve new markets or that use new technology to serve existing markets. When you are conducting market research to explore an opportunity for a new product that will serve your existing market, you are more likely to know who to interview and therefore may be able to get more helpful insights. Exploring new markets is more risky, because

The best market research may still be misleading.

you may be unable to identify the most important people to interview or sample. Or, you may ask misguided naïve questions and not be corrected by respondents.

One very successful materials company found that its successes outnumbered disappointments only when aimed at current customers and decision makers [28].

Industrial versus consumer markets illustrates another dimension of difference in detecting market opportunities. Marketing may be thought of as what you do to fill the factory and sales may be thought of as what you do to empty the factory. Industrial marketing departments are commonly staffed with promoted field sales personnel, customer service personnel (who are often trained as engineers), or other people lacking formal marketing training. They thus have a sales rather than a marketing orientation. Whereas all marketing people are sales oriented, many salespeople lack a marketing orientation. A good salesperson loves to sell $20 bills for $10 because it is easy to make the sale. A good marketing person is excited by the challenge

The sales orientation is to overcome objections, whereas the marketing orientation is to listen to objections to try to learn from these.

of trying to sell $5 bills for $10. They want to devise a bundle of valuable benefits that have low production and distribution costs or to devise packaging, advertising, or other promotional techniques that generate excitement and make it possible to sell a low-cost product at a high price.

Whirlpool placed prototypes of new washers and dryers in the homes of employees' mothers to test them and obtain users' reactions. This simple technique preserved confidentiality and assured candid feedback.

For many years, General Motors required its employees to rent only GM cars when traveling. This policy precluded obtaining widespread firsthand information about the features of competitors' products.

The availability of information is another distinction between business-to-business (industrial or commercial) and consumer markets. In the former, there are usually small handfuls of people who can provide essentially all the market research information you may require. The challenge is to identify and interrogate these few experts. Information is diffuse and widespread in consumer markets, with each person's opinion in the target segment having essentially equivalent value. Thus, the market research samples are relatively large, typically 1500 people, to obtain 3 percent accuracy.

Consumer markets require different market research techniques than business-to-business markets.

Other market stimuli may lead you astray. That is, you may get a very encouraging indication about an unserved market need, but the ability to capitalize on it may depend on infrastructure that does not exist or is marginal.

The attractiveness of a videophone depends on how many others are attached to the telephone system. If you are one of only a few owners of such a new product, it will have very limited utility. That initial limitation can be a fatal impediment to sales, regardless of the possible attractiveness of videophones.

Global warming and finite petroleum reserves provide a market stimulus to the development of electric vehicles. However, the attractiveness of an electric car depends, in part, on the number and convenience of recharging sites. This inherent limitation has stimulated the development of experimental electric cars employing solar panels [29].

When initiating the FFE interval because you have detected a market opportunity, it is important to identify the market segment that will benefit most from your new product. Do all the preliminary investigation of this segment. If you cannot serve it adequately, it is less likely that other segments, even if very large, will be well enough served to justify the effort. Consider, for illustration, a hypothetical new energy-saving switch that can sense automatically when the last person leaves a room and turn out the lights. If the use of this switch is not justified in new construction, it will be unlikely to make any inroads into the potentially larger building repair or remodeling market. In new construction, there is no existing switch, so something has to be purchased and installed. In existing construction, a workable nonautomatic switch is already installed.

The market stimulus—any stimulus, in fact—must be examined in light of the cost-effective benefits the prospective new product will provide to users. The user's costs include not just the purchase price, but also installation, training, operating, and maintenance costs. It may be helpful to consider whether you should provide a lease option rather than require only a sale for some new products.

The new GM EV1 electric cars being test marketed in southern California and Arizona were only available for lease in 1997 and 1998. This limitation was intended, in part, to assure that the vehicles would be used only where battery-recharging facilities were available and that the users clearly understood all the other various limitations of these early "green" models [30].

Strategy Stimulus

The FFE intervals for major drug companies are commonly stimulated by chosen strategies. These companies consciously select a few classes of disease upon which to concentrate. Given the vast array of diseases and potential genetic disorders in the world, there would otherwise be an almost unlimited range of avenues to pursue. Their early stage research is therefore confined to explorations for molecules that may be usefully active in these restricted categories, and this work, in turn, channels new product development into the strategically selected areas. Their business strategy drives their product development strategy.

Companies, even very large ones, must maintain a strategic focus and not try to everything.

From mid-1993 through the end of 1997, Eli Lilly reduced its R&D investigations from eight areas to five areas: the central nervous system and related diseases, infectious diseases, cancer, cardiovascular diseases, and women's health [31].

The concept of disease management originates with our customers. . . . the procedures go from prevention, to diagnosis, to treatment, to after care. . . . Since our customers are thinking this way, we needed to align our strategies to them—and we did. . . . we carefully select diseases on which to focus. . . . They also meet several other criteria—prevalence, seriousness of impact, and potential for more cost-effective treatment— that add up to an ability for Bard to make a meaningful contribution to the advancement of care, and a willingness of payers and patients to spend appropriately for care. . . . the disease states of vascular, oncology, and urology [have been] chosen. . . . We are no longer defined by our technologies, but by the problems that clinicians are trying to solve. Being closer to the customer's business gives us better insight into what the real problems are and what products and services are needed to manage these diseases [32].

Source:"Refocusing Bard's Business," reprinted with permission of C.R. Bard, Inc.

The important thing, however, is not how many products we have in the pipeline or how much we spend in research but rather the nature of the products we discover. That is why our research strategy is specific and focused. We are concentrating on disease states with large patient populations where we can apply new knowledge about pathways of disease to develop novel drugs [33].

Source: Chairman's Message to Stockholders, reprinted with the permission of Merck & Company.

A strategy stimulus—"we must be in this business"—may require the development of complementary infrastructure. This is, as we saw above, the case for electric vehicles that use rechargeable batteries. Today, car manufacturers are committed to trying to develop energy-efficient nonpolluting automobiles, and complementary infrastructure is required for many candidate technologies.

Another possible electric vehicle could run on very efficient fuel cells, which only produce water as a by-product. However, these cells require hydrogen for fuel, and today there is no complementary infrastructure either to produce this in adequate quantities or to safely store and distribute required quantities of this potentially explosive gas.

One approach to this challenge is to team with other companies. These may be the eventual adopters of the product, channel intermediaries, manufacturing partners, or sellers of complementary products or services.

3-D Systems, developer of the first stereolithography products that achieved widespread use, required both suitable liquid photopolymers and startup funding. They solved both problems by teaming with Ciba-Geigy, which made a major investment in return for an ownership share, and supplied the liquid photopolymers.

There are three other comments about opportunities detected by thinking about or considering a company's strategy. First, you may preclude or encourage licensing technology that is developed in other companies. Second, you may also preclude or encourage sales through your own sales force (or alternative channels). Third, clearly, the emergence of a "killer idea" can change the strategy for any company, although these ideas are uncommon.

Idea Stimulus

The flash of insight, the sudden recognition of a possibility, and the accidental or unintended observation of an opportunity are examples of an idea stimulus. The idea need not depend on new technology.

Many years ago a very clever engineer toured the Bureau of Engraving and Printing with his family during the spring school holidays. He observed dozens of people inspecting printed sheets of uncut currency notes to identify and mark defects. (This labor-intensive inspection was performed prior to cutting the bills to final size and was inherently somewhat idiosyncratic depending on the particular inspector and their alertness.) A defect was—and still is—the presence of ink where it should not have been or the absence of ink where it should have been. The objective of careful inspection was to assure very high quality and thus impose a barrier to easy counterfeiting. He realized that this very boring inspection task could be both automated and consistent, taking advantage of optical comparison techniques that were well known. It took several years to overcome problems that were not obvious and design a machine with fast enough throughput to be feasible economically.

Another approach is to stimulate ideation. There are many techniques by which this can be done. Brainstorming, conventional or facilitated, is a well-known example. Today, there is also a variety of software that is intended to stimulate ideation [34]. If you lack any ideas or feel that the other three means to detect opportunities have not produced an attractive set of options, you may wish to stimulate ideation by one of these techniques. Some elements that can help promote idea generation include:

- Consistent support by senior executives and managers
- A process that is simple and fair (in which a less good idea, not the proponent of the idea, is rejected)
- Some kind of reward system, either extrinsic or intrinsic or a combination

To take advantage of external knowledge and needs, as a first step, perhaps conduct approximately two dozen telephone interviews with people to try to discover some clues about opportunities. Or try living with potential customers and users in your existing market or chosen target market.

A company supplying equipment for trucks was not regarded as the innovative leader for its particular class of products. A key competitor was being invited to sit with fleet owners and truck manufacturers to discuss future needs and join in the development teams that were being formed to create future new products. Key executives at this equipment supply company insisted that they fully understood the market environment. They refused to put their marketing and engineering personnel in truck cabs and distribution facilities as observers where they might have a sudden insight. As time went by, they continued to lose market share because they could not identify any ideas for innovative products.

It is, of course, necessary to really understand the intended market, the channels of distribution, and purchase patterns to be successful. Some very attractive ideas that offer great promise run afoul of a superficial appeal that misses market realities.

A hypodermic needle that essentially prevents accidental injury is a very attractive idea, and a retractable needle is not difficult to design. The resulting product, however, is finding only limited appeal because most

(continued)

> major hospitals have long-term buying contracts awarded to low-cost suppliers. These suppliers are reluctant to buy the better but higher-priced product. The higher cost will diminish the suppliers' profits and the system benefits of fewer injuries do not accrue to the suppliers [35].

A final consideration is how many ideas it requires to obtain a commercial success. For instance, the vast majority of new medicines for which clinical trials are initiated do not enter the market successfully [36]. There are two problems with the vast literature on the topic of new product idea mortality: The first is deciding what constitutes an idea at the beginning; and the second is the definition of a commercial success. One recent article claims that 3000 ideas are required to produce a commercial success [37]. Another researcher states that 11 new product ideas may be required to generate one successful new product [38]. Recent research conducted by the Product Development & Management Association finds that only seven are required [39]. Clearly, there is a vast difference in these findings, but what is unmistakable is that there is an idea mortality between early enunciation and subsequent commercial success. You will need several or many ideas if you wish to have a commercially successful result.

Very few exciting or promising new product development ideas lead to commercial success.

Hybrid Situations

Hybrid situations are those where there is no single initial stimulus (technology, market, strategy, or idea) that is clearly apparent to outsiders. In fact, there may be a mixture for which even employees who are intimately involved may be unable to isolate the precipitating stimulus.

> Diebold Company potentially furnishes an example (at least to this outsider). This fine company dominates the market for automated teller machines. In recent years, facing an industry slowdown, they have diversified to provide medication-dispensing systems that make use of much of the same technical expertise. Other diversification is aimed at providing security systems for campuses, libraries, and arenas. Which of the four stimuli drives these imaginative endeavors is unclear, but the efforts seem to make good sense.

SHORTENING THE FFE

Four stimuli to initiate the FFE interval were discussed In the preceding sections. Now it is time to consider what else must occur during the FFE interval. To shorten the time-to-market or time-to-profit, you may wish to shorten the FFE interval. As shown in Figure 1-7, this may not be necessary if you can gain time elsewhere, but it is often desirable.

> **Keep your eye on the goal, which is getting to the profit objective fast.**

Fundamentally, the FFE interval requirement is to do enough work in marketing, technology, and manufacturing to assure that there is an attractive business case for the putative new product and to preclude an unrealistic S&G interval development schedule. Therefore, there can be five elements that your firm may wish to include in its fuzzy front end practices:

1. Business strategy formulation
2. Product strategy formulation
3. Idea generation
4. Idea screening
5. Business case validation

The first two elements are concerned with selecting areas in which to seek opportunities, which is somewhat analogous to deciding where to plant seeds and what kind of fertilizer to use. In a sense, the notion is to find important problems that are worth solving.

The last three elements deal with the process of discovery and innovation. This is not merely a random process of generating ideas that are exciting to your marketing, technical, or scientific talent. Rather, the company's business and product strategy guide the process, and ideas are sought and moved forward only because they support these strategies. Opportunities—regardless of the stimulus—are evaluated carefully and objectively so that the need is well understood. This discipline is in contrast to the image of the chaotic genius, who is inspired by random events to create the unusual.

> **There are more good ideas to pursue than resources to advance them.**

For example, idea generation includes not merely generating and recording ideas, but also involves enhancing the ideas. Similarly, screening includes judging the ideas objectively against strategy and then deciding the priority of each promising idea.

What's the bottom line? New product development professionals have developed some sensible structures for the FFE interval. This structure provides a way to move more systemati-

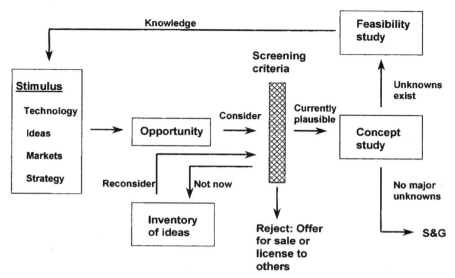

Figure 2-8. One approach to a systematic process for the FFE interval.

cally and rapidly through this initial group of new product development activities. If your company has not yet introduced a similar structure for the FFE interval, you may wish to take advantage of the emerging best practices.

Figure 2-8 illustrates one possible set of FFE interval activities. This is a "busy" figure and is best understood by going through it in detail, beginning at the left:

· Business and product strategies determine where we look to detect opportunities, although the opportunities we wish to pursue may originate as a result of any of the four stimuli discussed previously.

· Raw opportunities are considered systematically, briefly and quickly screened, and sorted into three categories:

1. Reject as not suitable for our business in the foreseeable future. These may be offered for sale or license to others.

2. Suitable but for which adequate resources are not currently available. These are deferred and stored in an "idea bank" or similar repository and are reconsidered periodically.

3. Plausible and resources are available to conduct a concept study.

· The goal of the concept study is to determine if there are any major unknowns about the market, technology, or manufacturing process. If there are major unknowns, no development schedule will be credible, so a feasibility study is appropriate.

- The objective of a feasibility study is to create knowledge, not necessarily a new product. The purpose is to eliminate unknowns, so the feasibility study must be thorough but should not be leisurely. The resulting knowledge may produce additional new opportunities or lead to termination of the investigation.
- If there are no major unknowns and the business case is satisfactory, the product concept is a candidate for development and launch and the company is ready to initiate the S&G interval.

The concept for an "electronic nose" provides an illustration of a product development effort that is still in the FFE [40]. As described at the time of the news report, the developers have conducted promising experiments but had yet to build a prototype. Sensitivity to specific smells is therefore probably uncertain. Durability and many other factors must presumably be evaluated. Market size depends on the device's cost, which is presently unknown and probably depends on production volume. In short, all the issues that should be addressed in the FFE are still awaiting resolution.

In some cases you can arbitrarily set a limit on the FFE duration, as, for instance, Fluke Corporation has done in the case cited earlier in this chapter. The advantage of a "sunset" date is to assure that the FFE work is carried out with a sense of urgency. The disadvantage of an arbitrary sunset date is that important work may be done superficially or not done at all, which will create subsequent problems in the S&G or PPS intervals.

A multifunctional team must carry out the FFE interval process illustrated in Figure 2-8 (or any alternative FFE interval process that your firm adopts). The team (i.e., the project's human resources) must be multifunctional to avoid tribalism. (Many writers use the term *cross-functional* rather than *multifunctional*. But *cross-functional* has a connotation of crossed wires, which *multifunctional* does not have. It is hard enough to reduce the tribal loyalties of departments and other functions, so the choice of word can be important, and I use *multifunctional* exclusively.)

There is "no free lunch"; resources must be committed if you want results.

Although primarily only marketing and technology personnel (or even just one person from each function) may carry out the initial FFE interval work, later work (especially the development and proof of the business case) must involve all functions. It may be helpful to realize that this approach exploits the firm's core

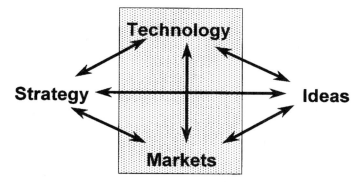

Figure 2-9. *Core competencies.*

competencies and nicely ties in with the model of four stimuli, as shown in Figure 2-9.

One of the key principles for this FFE interval is to preclude the necessity to "invent on schedule." That is, we want to initiate the S&G interval with proven technology. (In those cases where it is a business imperative to "bet the company," sometimes the gamble is made to proceed with unproven technology. This does happen occasionally and may be the lesser evil, as long as it is recognized that this is a very high risk approach and not something for which a development and launch schedule can be guaranteed.)

If you start the S&G interval with major unknowns that have not been removed during the FFE interval, your launch schedule will not be credible.

Another key principle is to try to assure that the product that you intend to develop will be adequately profitable when it is produced. This desired profitability obviously is a critical contributor to achieving the profit objective quickly. Assuring that your product's gross margin (or difference between selling price and factory cost) is suitable is one aspect of this. But gross margin depends on a very complicated mix of disparate elements and these must be carefully considered—if not fully resolved—during the FFE interval. Figure 2-10 illustrates some of these interacting elements:

- The buyer's benefit-to-cost determination—their propensity to purchase—depends on the product's promised (or advertised) performance, the date it will be available in sufficient quantity, and the selling price. In broad generality, more performance, earlier (but not premature) availability, and lower price produce more demand and lead to more sales volume.

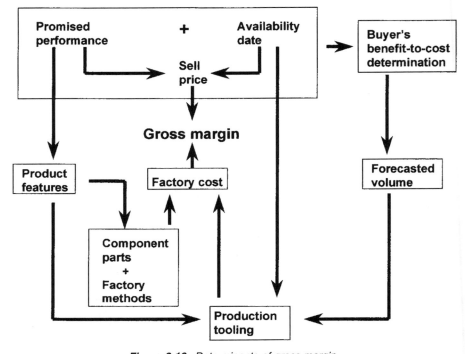

Figure 2-10. Determinants of gross margin.

· The product's promised features depend on the selection and realized performance of component parts.

· The selection of production tooling (and its cost, both capital and expense) depends on the product's features, the date of initial shipments, the rate of production volume buildup, and the forecasted ultimate volume.

· The factory cost will depend on the costs of parts and components, the choice of factory methods, and the design and operating cost of the production tooling.

· The difference between the selling price and the factory cost is the gross margin. This is determined largely by choices made during the FFE interval, well before production begins.

Other elements such as labor and overhead costs, depreciation, or utility costs may be significant. If so, these must be considered. In the case of many products intended for consumer markets, emotion overrides a strict benefit-to-cost determination.

Many companies take advantage of activity-based costing (ABC) principles to better understand the true costs of a new (or other) product [41]. This is not an easy technique to implement, but may well be highly desirable to facilitate understanding of the profitability of any product.

CLEAR REQUIREMENTS

Figure 2-11 illustrates many of the varied factors that dictate the requirements a new product, and these requirements are what must be dealt with and satisfied by the specification. Although customers, users, and channel intermediaries (such as wholesalers, distributors, and retail outlets) are commonly thought to be the key drivers for a new product's requirements, other factors are also critical. Company standards may include varied factors, such as a minimal acceptable financial return, quality levels that reduce liability risks, and compliance with company values. You must also be certain that product design does not

Be certain that the requirements and specifications derived from them, when satisfied, will solve real problems for key customers and users.

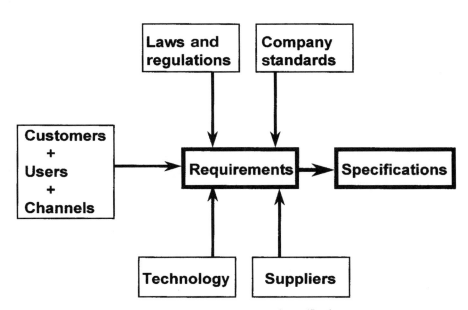

Figure 2-11. Requirements and specifications

ignore aspects of company strategy. For instance, for a company goal to be an international supplier was undercut by a product control panel that was entirely in English (and without any international symbol icons).

SUMMARY OF KEY POINTS

· The FFE may be initially stimulated by strategy, ideas, technology, or markets.

· You must deal with and assure consistent resolution of all four factors during the FFE.

· A successful FFE effort will remove unknowns about the market, technology, and production processes so that breakthroughs or inventions are not required during the S&G interval.

· Although there are initial uncertainties about issues that must be resolved during the FFE, a schedule plan and sense of urgency are helpful.

· Clear product requirements should be produced at the end of the FFE.

3

The Stages and
Gates Interval

INTRODUCTION

The stages and gates (S&G) interval starts when:

- Enough is known about the opportunity and requirements, so that the goal for the S&G interval is clear.
- The concept no longer contains major obvious unknowns about the market, technology, or production process.
- A plausible business case for the product has been constructed.

To put this differently, a key FFE interval objective was to set the S&G interval's goal before an ill-considered or marginal idea is developed—typically at high expense—into finished form and launched into the market.

The S&G interval ends when products that can be offered for sale are shipped. Some companies mistakenly believe that shipment is the same thing as commercial success. Although you clearly must start to ship a product to be commercially successful, more is involved. Actually, successful commercialization is really the end of the PPS interval.

S&G INTERVAL INGREDIENTS

Historically, what I call the S&G interval has been synonymous for many people with the entire new product development (NPD) process. To be specific, many people did—and some still do—believe that NPD starts with either a brilliant inspiration (the light bulb going on, as illustrated in Figure 1-4) or an agreement on the specification and ends with a new product's launch. Others see NPD as project management, typically the sole responsibility of the technical function (i.e., research, development, or engineering). However, as I have already explained, the S&G interval is only a portion of the entire NPD process. Specifically, the total NPD process is composed of the FFE, S&G, PPS, and CS intervals, as shown in Figure 1-5.

> *Successful NPD involves more than just managing technology and it is not the sole province of the R&D or engineering functions.*

> The S&G interval is that portion of Baxter Healthcare's RIGHT–Products–RIGHT slogan that is concerned with "Do the Products Right." That is, correctly execute the design, development, testing, qualification, and launch of products—and do so quickly.

Most companies use a fast stage-and-gate process to manage their NPD efforts, especially the FFE and S&G intervals. Introducing a new product—even one that has been nurtured and vetted through a successful FFE interval—requires a series of actions during the S&G interval.

Figure 3-1 shows the efforts in the S&G interval (set specification, time-critical design and development, and continuous improvement) and the FFE and PPS intervals that precede and follow. The reason to show these other

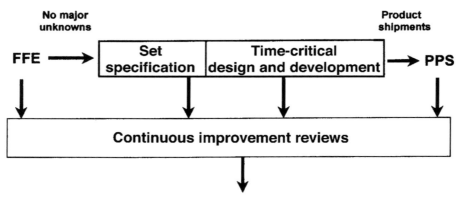

Figure 3-1. *Elements in the S&G interval.*

activities is that many of the resources required for the S&G interval are also involved in them. Your firm's resource allocation (or lack thereof) has important implications for how quickly you can launch the product or achieve your profit objective, as I discuss later.

A number of years ago, a device manufacturing company found that the market it had created for a weatherproof electrical connector was being taken away by new competitors who were offering lower-cost connectors. This company had initiated a NPD effort that was intended to reduce the cost of its connector. The team was already engaged in S&G interval activities, but this development effort was not yielding promising results, and the team kept oscillating between setting revised specifications and attempting time-critical design and development (TCD&D). Market research was then initiated, and this revealed that a new but improved application tool could substantially shorten the amount of time a technician required for field installation of the device. Thus, when the new tool (which was unique and specific to the company's connector) was developed, it could save the user enough technician labor cost during installation to overcome the higher connector cost. This company had been engaged (unsuccessfully, as it happened) in trying to develop the wrong product. Going back to the FFE interval to select the right product led to a revised S&G interval effort to develop a successful new installation tool rather than a lower-cost connector.

The S&G interval is only one part of the total NPD process.

The apparently elementary graphs in Figure 3-2 depict the key relationships that affect speed during the TCD&D interval. Although deceptively simple, this is really very important and can provide significant insight. This shows the critical S&G interval trade-off. Figure 3-2*a* illustrates that higher levels of specification difficulty lead to a longer development time. (At very high performance levels, no one knows how to develop the product, and the schedule time becomes infinite.) Figure 3-2*b* shows that the availability and use of more effective resources can alter the performance-schedule relationship, so that more effective resources can achieve a particular level of specification difficulty on a shorter development schedule. Figure 3-2*c*

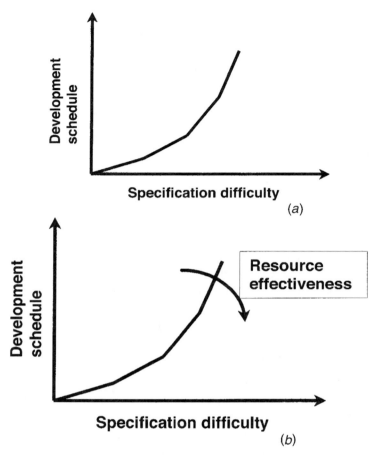

Figure 3-2. *(a)* Relationship between specification difficulty and development schedule; *(b)* impact of more effective resources; *(c)* specific development schedule dictated by a specific specification and level of resource effectiveness; *(d)* development schedule illustrated in Figure 3-2c shortened by more effective resources; *(e)* general relationship and implications for shortening development schedule.

Time-to-market can be shortened if the specification is simplified or if more effective resources are applied to the TCD&D effort.

illustrates that a specific level of performance and resource effectiveness determines a specific time-to-market. In general, a product with a more difficult specification normally has a longer time-to-market (i.e., launch) than one with an easier specification.

However, at any level of specification difficulty, making use of more effective resources, as shown in Figure 3-2d, can shorten the time-to-market. The converse is also true, that use of less effective resources

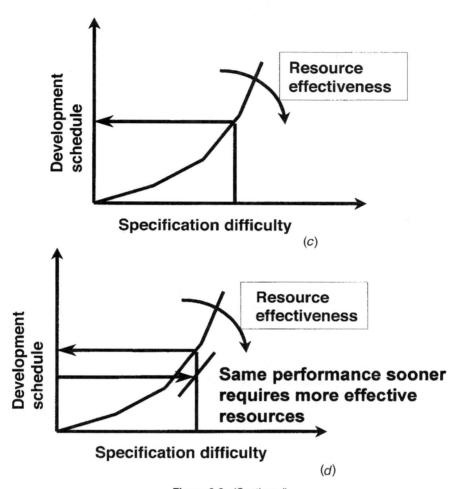

Figure 3-2. (Continued)

causes a longer time-to-market. Thus, to shorten time-to-market, you must aim for simpler specifications or plan to utilize more effective resources (either people or equipment), as summarized in Figure 3-2e.

There are many ways to improve the effectiveness of resources. Some examples include:

· Use employees with successful prior experience.
· Use employees who are proficient for their required tasks.
· Assign full-time dedicated personnel for longer or more critical tasks.
· Assure that facilities and personnel are available when needed.

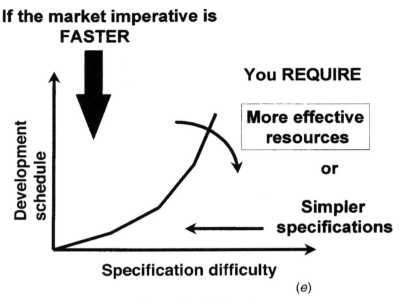

If the market imperative is FASTER

You REQUIRE

More effective resources

or

Simpler specifications

Development schedule

Specification difficulty

(e)

Figure 3-2. (Continued)

Obviously, these actions are not always possible, and other options may be appropriate in a particular company's culture.

In the following sections we provide more detail about the three key ingredients of the S&G interval: the specification, time-critical design and development, and continuous learning (which is covered more extensively in Chapter 6).

The Specification

It is critical to set a firm specification early if you wish to get to market quickly. Although the requirements should have been clearly defined in the FFE, the detailed specifications that will satisfy the requirements must now be determined. However, there is a trade-off, as shown in Figure 3-3. If you set the specification quickly but prematurely, meeting the specification may be inordinately difficult and time consuming. Taking a very long time to analyze options, consider trade-offs, and establish a firm specification can lead to "analysis paralysis," and this may then become the pacing item. There is not an analytic way to find the minimum total time, so taking advantage of the experience of senior product developers is usually the best you can do in determining when you have done enough specification adjustment.

Do not spend either too little or too much time agreeing on the specification.

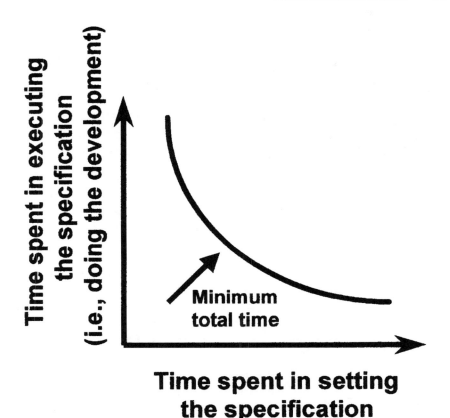

Figure 3-3. *Minimizing the time-to-specification.*

As discussed in Chapter 2, the concept of a family of new products is clearly helpful when there are many possible specification options. Although you should obviously have an initial product that is easy to produce fairly quickly (even if its general utility is somewhat limited), other variations can either be launched simultaneously or follow rapidly.

The absence of a firm specification can promote work being done at cross-purposes. In some development efforts (especially those that may be inherently lengthy), an incremental freeze (or "creeping freeze") is both necessary and effective.

In the case of some electromechanical products, the exterior styling may require several design iterations and perhaps buyer or user market research (e.g., with focus groups). In such situations, it is often helpful to reach an early agreement on the mounting points to which the "guts" of the

Developing a new product to satisfy a specification is similar to walking on water—It's easier if it's frozen.

Example

Case shape not
settled

Keypad layout settled

Electronic innards
must be rushed to
completion

So, define keypad-to-
case interface
mounts

Figure 3-4. *Incremental freeze.*

device will attach, as illustrated in Figure 3-4. When this approach is taken, the case design is limited only by the location of mounting points (and, for the example illustrated, the visible portion of the keypad). In the example,

Freeze as much of the specification as you can as early as possible.

the exterior case and the working parts inside can be developed in parallel, thus shortening the total S&G interval.

Specifications can involve both objective (quantitative) and subjective (qualitative) elements. Examples of objective elements include the cost of the product's ingredients, package size, weight, various functions (e.g., speed, resolution), and similar items. Almost all companies do a fine job with nailing down the objective elements of the specification, although there are occasional exceptions.

Objective specifications should be set explicitly.

A company was developing a wheeled stretcher for use in ambulances and hospitals. One element of the specification was that the stretcher should be "stable." That is, it should not tip over when a heavy patient was being transported. No one took the time—or realized the important need—to convert this to an explicit objective specification. Stipulating the maximum patient weight, the maximum stretcher elevation, and the maximum lateral force it had to withstand before tipping could have done this. The initial prototype—developed with considerable time and cost investment—tipped over easily and was clearly unsatisfactory when first tested.

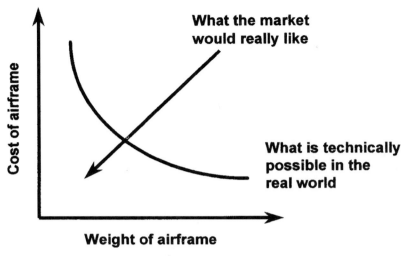

Figure 3-5. *Specification trade-off.*

Figure 3-5 illustrates that technical trade-offs are frequently required when setting objective specifications. This is never easy, because honest people will often disagree on the best way to compromise between conflicting objectives. Quality function deployment (QFD) can be helpful for this [1]. However, the full use of QFD is very time consuming and most companies find that a simplified form of QFD can be used effectively to make and record these design trade-offs. A subsidiary benefit of any form of QFD is the emergence of a common view in the multifunctional team of critical design factors and key trade-offs.

Notebook computers provide a commonly encountered illustration of specification trade-offs. A comfortable and fully featured keyboard limits the extent to which size can be reduced. A bright display screen and a faster microprocessor both require more power, which requires larger and heavier batteries that increase weight. Thus, low weight—which is highly desired—is hard to achieve with current technology, and each manufacturer makes varied compromises among these features [2].

Testing, especially during the S&G interval, is both the means by which achievement of specifications is demonstrated and also how problems in manufacturing and customer use may be reduced. Testing takes time. How much is too much? How much is enough? At what point is too little testing done and

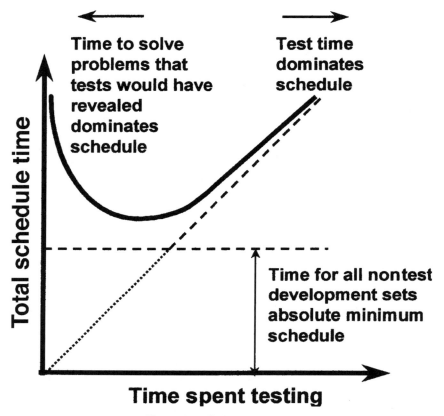

Figure 3-6. Optimum test time.

the risk of problems unwise? Figure 3-6 illustrates that if you do not do enough testing, problems that might have been detected are not. These problems will not magically disappear just because you did not look for them. They will arise later, causing problems that increase your product's cost or delay your manufacturing operations—or both. Or, worse, the undetected problem may cause difficulty for a user, and, perhaps, a warranty claim.

If, instead, you do every test that anyone can plausibly propose, the time for testing will quickly become excessive. However, arbitrarily limiting the number of tests or the time for testing is obviously risky. What do you do if there is an anomalous result? Do you take extra time to repeat tests or perform extra tests to clarify the situation? Or do you move ahead—as they did with the Hubble telescope mirror—to maintain the schedule? The bottom line is that you must use judgment and be flexible to find the appropriate balance.

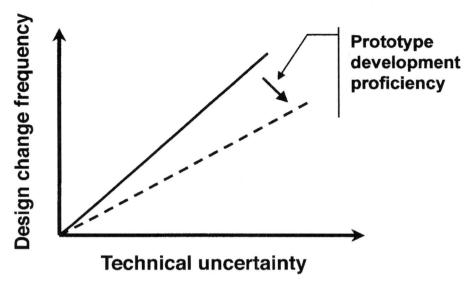

Figure 3-7. *Relationship between technical uncertainty and design change frequency.*

Testing of prototypes is clearly an effective means to reduce surprises and any design changes subsequently required. Therefore, a company that can quickly design, build, and test prototypes (including models, breadboards, and similar pre-product devices) should be better able to cope with the inevitable technical uncertainties in the S&G interval, as illustrated in Figure 3-7 [3].

Subjective (qualitative) element specifications and testing are another matter. Examples of subjective elements include styling and appearance, sensory aspects (e.g., taste, smell, sound), and so on. Most companies, with the notable exception of many food companies, do a marginal or poor job of specifying the subjective elements. Words like *attractive* are frequently used to describe the desired product styling or industrial design. The obvious problem is how to decide when this specification is met.

A very large corporation was engaged in a massive project to develop a new vehicle to be operated by a single user for moving earth. One element of the specification was that its design must be "contemporary." When queried about how this would be judged, the team was silent, as they realized that they had not yet grappled with the criteria for that decision.

Clarify how subjective specifications will be judged before much development work is done.

Unless the method and criteria by which a subjective specification will be judged is clearly delineated, developers can easily end up working at cross-purposes and time is often lost. In a small private company the owner typically makes the decision, so he or she must be kept informed. In a large public company it can be harder to identify a single decision maker, but this is what you must try to do. When a subjective specification element is required, it is crucial to clarify in advance how compliance will be judged.

A food products company typically stipulated that a prespecified fraction of a consumer taste panel must give the new product a prespecified (or higher) score on a blind taste test; for example, 8 of 10 must score it a 6 or 7 on a 7-point scale. Based on a long history, taste panel scores could be related to future market share. In addition, there would be explicit criteria for where and how the consumer taste panel was to be recruited: for example, a particular type of shopping mall (e.g., upscale or other) in prechosen cities that are chosen to be representative of the expected market for the new food product. The clear description of how a new product would be qualitatively evaluated avoided recriminations because the standards were quantitatively explicit and provided an assurance that a satisfactory share of market would be obtained.

Beta testing in which a product is evaluated by prospective users can be used to assess both quantitative and qualitative aspects of a product. The beta test product is generally considered to be essentially "right" by the producer, representative of what will soon be produced in volume, although it may have unfinished aspects. (However, for mass market computer software products, the beta test product may be quite unsatisfactory, and testing is generally confined to computer-literate "techies" with idle time and a willingness to explore developmental software.)

In the "good old days" when new products were scarcer and demand was almost insatiable (e.g., in the years following World War II), many companies set their price as a markup on their cost. They aimed at a specified set of product features and did not spend a lot of time or effort in trying to limit the product's production cost. The markup was chosen to be high enough to produce the company's desired gross margin above factory cost. Some companies still operate in this mode. Today, however, it is more common to design to cost, adjusting the product's features and benefits so that the

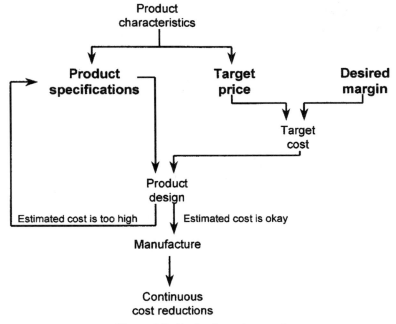

Figure 3-8. *Design-to-cost concept.*

selling price becomes a critical design target. To put this differently, the price is held sacrosanct while product features are adjusted to provide a best fit to some identified market opportunity. This concept is illustrated in Figure 3-8.

Time-Critical New Product Development

There is a vast array of tools that companies use to improve elements of the product development process or shorten the time devoted to some aspect of it. Many of these are particularly applicable to the S&G interval. Examples include QFD, computer-aided design (CAD), computer-aided engineering (CAE), computer-aided software engineering (CASE), computer-aided manufacturing (CAM), design for manufacturing and assembly (DFMA), stereolithography, and so on. Many techniques or practices have also become common, including six-sigma design [4], the statistical design of experiments, alpha and beta testing, and so on. All of these—and many others—have very useful roles to play.

Project management tools and techniques are also valuable but profoundly different than the array of tools just cited. This book is about shortening the time-to-profit, and the central value of project management tools is to help assure that any proposed schedule can be met. Schedules obviously depend—in part—on the array of design tools employed, the skill of developers in using these tools, and tests that are performed. Schedules also are critically dependent on the timely availability of appropriate resources, both human and physical. Project management software (or some other device to manage schedules and resources) is really essential because it can be used to plan a feasible schedule and identify the requirement for and availability of resources.

The resources that are actually available to the development team determine whether the schedule is realistic to complete the required work.

The most effective use of project management tools and techniques during the FFE interval and during the course of firmly setting the specification is to define, plan, monitor, and rapidly complete integrated groups of short-range tasks. These same project management tools and techniques are invaluable—and frequently essential—during the relatively longer duration of the TCD&D activity. In general, many people are now at work simultaneously on varied tasks, perhaps at separate sites or installing very expensive production equipment, in what must be a coordinated TCD&D effort that is frequently carried out under extremely tight and challenging time constraints.

Some people believe that NPD project goals are broad, complex, and unclear; the experience base is low; teams are multifunctional; management environment is chaotic; and the success rate is low. Thus, they believe that traditional project management techniques are ineffective, since these are typically applied to construction that is fixing a defined problem with a few specific goals, for which there is high experience, teams with a common background, a stable management environment, and a high success rate. However, the vagueness that such people ascribe to the goals of a NPD effort do not apply after the specification is set, and forgoing the use of project management tools and techniques at this stage is unwise [5].

It is relatively easy to establish project plans for the S&G interval after specification is firmly set. In fact, I believe that the inability to do this is a strong signal that the development team does not know how to accomplish the TCD& D work and launch the new product on the desired schedule.

The absence of a project plan and realistic schedule the for the TCD&D work is a danger signal.

Companies as varied as manufacturers of medical diagnostic systems, telecommunications devices, and brake components for tractor–trailer truck rigs have found that they can construct general templates for their typical NPD efforts.

In each case, the templates covered the entire FFE and S&G intervals and included a few PPS tasks. These templates each turned out to require about 500 separate tasks that could be identified and linked sequentially to predecessors and successors, and all these tasks could be separated into the discrete stages that were appropriate for each company. These data could be entered into project management software so that there was then a general template that served as a starting point for each subsequent NPD effort. The estimated schedule duration of each task obviously depends on its specific complexity and the resources devoted to it. In a particular NPD project, for example, a task would be longer if a less well-qualified person was responsible for it than if a more qualified person was responsible for it.

Companies commonly divide the S&G interval into varied stages. These stages include generic groups of tasks related to developing the new product and scaling it up for routine production [6]. Examples of stages include detail design, design demonstration, engineering model or prototype, system verification, manufacturing verification, and pilot production. Different companies—and even divisions of the same company—have idiosyncratic stage names. A myriad of tasks is included within a stage. Examples of tasks or activities might include system design, subsystem breadboard, test planning, component or subsystem testing, user documentation, training materials, and sales support. These activities—and dozens of others that are typically required—are sorted into the separate stages. Companies have as few as two or three stages in the S&G interval, and in a few cases, as many as half a dozen or eight. The stage and task labels would vary for a service or a purely software product, but the kinds of things that must be done are similar.

There is much similarity in best practices, but no two companies have identical development tasks or stages into which these tasks are separated.

A review of the S&G interval effort is carried out at the end of each stage, and these are called gate reviews (or stage–gate reviews). These gates, which require that cognizant executives or managers make a decision, are often labeled with "Go or Kill" or "Go, Redirect, or Kill." However, kill has a pejorative connotation, implying failure. Thus, there can be an inherent reluctance to stop a development effort. Regrettably, in a few companies some NPD projects that have really been abandoned persist and absorb some resources that might be more usefully applied elsewhere. A culture change is required in which stopping a development effort is seen as an opportunity to learn how to do better in the future. One option may be to label the "Kill" or "Stop" decision as "Exit and disseminate new knowledge." Although this is a mouthful, it may promote a shift to a more effective way of behaving. This is the basis for continuous learning, which I discuss later.

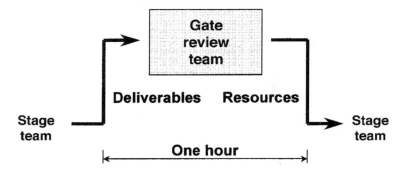

"Go" unless:
Deliverables from prior stage are missing
Business case has become marginal
Resources can be used more effectively
on a higher-priority assignment

Figure 3-9. Concept for the stage–gate review.

> *Every discontinued NPD effort can provide a learning experience if the lessons are proactively sought and widely disseminated.*

There are many potential problems with gate reviews. Some examples include overelaboration, exclusively rigid gate options, the "blame game" ("I gotcha"), inappropriate reviewers, and concentration on unimportant issues. The blame game occurs when a reviewer, typically a manager from a higher level than anyone on the development team, chooses to prove his or her brilliance or wisdom and destructively shoots holes in some aspect of the development team's work. (Obviously, if there is some undetected flaw, it should be addressed and resolved, but the manner in which this occurs and the tone of the discussion can be helpful or destructive.) A related problem is a managerially imposed specification alteration or project redirection that is suddenly injected, typically making the goal more difficult without recognition—or acknowledgment—of the schedule consequences.

Figure 3-9 illustrates an efficient and fast way to conduct stage–gate reviews. The multifunctional project team is expected to have completed all work in the prior stage and deliver whatever is due (e.g., market research, design drawings, test results, manufacturing plans), depending on the stage and the detail of a company's NPD process. If the pertinent management insists on confining the review to one hour and demanding that it be informal (but organized), the tendency to stage a very fancy "dog and pony" presentation with elaborate graphics and a carefully rehearsed script can be reduced. The notion is basically to leave the recommendation to continue or discon-

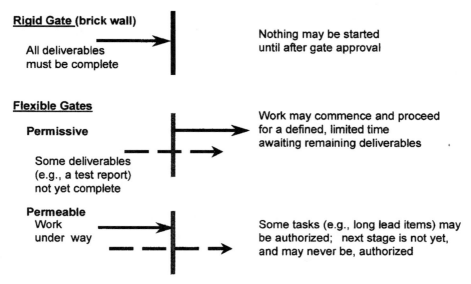

Figure 3-10. Gate options.

tinue the project in the hands of the development team. The team that has completed a stage successfully should also provide their answers to the obvious key concerns about the next stage: the biggest issues to be resolved, specific resource requirements for the proposed schedule; and the deliverables. Management should only stop or suspend the development effort if there is clearly a higher-priority assignment for the required resources.

Figure 3-10 depicts two forms of flexible gates that can be used to increase the gate decision options. These can shorten time-to-market during the S&G interval. However, replacing rigid gates with permeable and permissive gates imposes some increase in risk. Nevertheless, on balance, using flexible gates is often the lesser evil.

There may also be inappropriate reviewers. Senior executives should not attend some reviews. Some companies have NPD processes with gates that are primarily design review validations, and these are not likely to make effective use of a senior executive's time, so attendance should be delegated to others. This reduces delays due to the frequent unavailability of very busy executives. Such delays can be eliminated if scheduled reviews are given schedule priority (over all matters except true emergencies) and are put on executives' calendars well before the due date.

Another problem with stage–gate reviews is the tendency of some reviewers to concentrate on the actual versus planned development cost. The actual *development cost* is relatively unimportant per se, after the financial payback has been calculated and justified. The critical management issue is more commonly the allocation of a company's limited resources (principally human, but also physical) among competing NPD investment (and other) opportunities. It is more important for senior executives to ask whether the continued investment in a particular NPD project is the best (or, at least, justified) use of the firm's inherently limited resources and not to focus on whether the NPD project is over or under budget.

Do not focus on NPD development cost during reviews.

Two other topics deserve attention at this point: how to coordinate the resources on a single project and on multiple projects, and how to avoid some of the practical problems that are frequently encountered in the S&G interval, especially the TCD&D portion.

Resource Coordination   Timely and cost-effective NPD inherently requires a multifunctional project team in which many tasks are carried out simultaneously. There are often problems achieving this because there are different orientations among the people in different functions. Take the trivial example of development engineering and manufacturing. Most manufacturing departments are faulted for excess scrap, but every new product that enters manufacturing initially creates greater levels of scrap. Or consider the reluctance of a chemical production facility that is operating around the clock to interrupt production of revenue-producing shipments to conduct trial runs of a new formulation that is certain to be initially troublesome.

In a major chemical company, new chemical formulations progress from bench-scale development to small batch development to semiproduction environments. Ultimately, a new product—if it is to be produced in volume—must enter production. At this point it must compete with the internal plans of the production facility, which may include scheduled turnarounds (converting from one regular product to another) as well as "bread and butter" production of products for revenue-producing sale. The production facility will also give priority attention to resolving any production upsets, such as out-of-tolerance product.

Similar problems have been observed in a company producing wire and cable, where production machinery is operated around the clock and seven days per week. Obtaining production downtime for conducting the required trial runs of a new product is—whether agreed to or not—of lower priority than continuing the production operations. In addition,

(continued)

such new product trials inevitably produce scrap, which leads to an inherently undesirable cost variance, which is normally charged to the production facility's overhead budget. Finally, any production facility of this sort commonly may suffer from a not-invented-here (NIH) attitude.

Different problems may bedevil relations between the marketing and operations functions. One example is marketing's pressure for a faster ramp-up to full volume shipments. Similarly, marketing and the technical functions often dispute the relative importance of product features and cost and the schedule for their realization. Coordination and multifunctional cooperation are therefore crucial. A critical path schedule employing time-scaled tasks that also displays explicit task interdependency can greatly encourage this teamwork [7].

Microsoft Project has emerged as a very commonly used project management software package for NPD, at least for all but massive undertakings (such as a new commercial jet airplane). It is possible to easily include other useful information in the tabular listing that accompanies the Microsoft Project Gantt chart. Figure 3-11, a Gantt format created with Microsoft Project, illustrates this concept. Key data can be listed in the optional user-labeled columns that normally appear on the left side of the Gantt chart bars. Done correctly, the specific nature of each function's responsibility for and involvement in each task can be indicated clearly by the use of such a schedule format. This important information previously was embodied in separate coordination and task responsibility (or accountability) matrices [8]. In the case of Microsoft Project, columns for each organizational function can be used for each task to indicate the nature of their responsibility in that task as accountable (A), concurrence required (C), or input required (I). Other microcomputer-based software for project management may also prove to be suitable if there are user-defined data options for each task.

Project management software is very valuable to help you coordinate the work of resources from varied functions on a given NPD effort and to identify overcommitted resources across all efforts.

Many project management software packages allow the forecasted resource usage for given dates or time periods to be summed for several projects. Microsoft Project is one example of such software (although the version that is current as this is being written may not be the best for this specific chore). Using this capability, it is possible to discover if a company's development resources are or will be overcommitted at any particular time. If this is the case, one or more development pro-

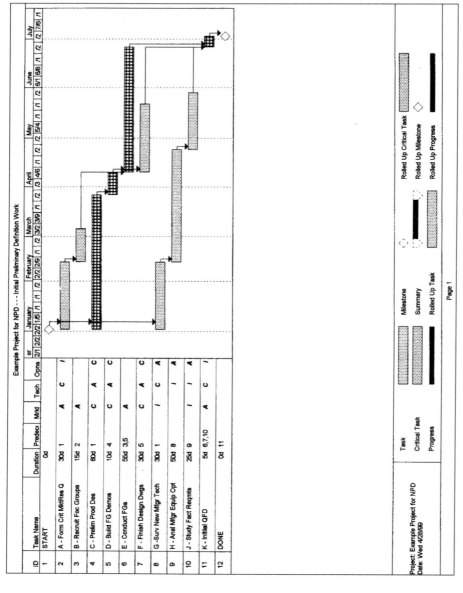

Figure 3-11. Project management software can be used to indicate the role of each function in each development task. (A denotes accountable, C denotes concurrence required, and I denotes input required. There must be an accountable function for each task.)

jects will be delayed—people or equipment cannot effectively do two things at one time. Learning of this conflict before it occurs allows senior management to explore alternatives and replan as required.

Some Practical Problems to Anticipate Once you are engaged in the S&G interval, your firm's "burn rate" of money is likely to be high. Although delays in either the FFE or S&G intervals will obviously delay the new product's launch, the expense and capital cost consequences of delay in the S&G interval are much greater than in the FFE interval. Here are a few real (but slightly disguised) quotes from product development managers and engineers.

> "My manager insisted that I use a partially idle engineer to work on the development project. While this person was eager to learn, the engineer really lacked the necessary skills to perform the immediately required tasks. The result was that I had to devote considerable time to provide assistance or redo the engineer's work."

> "Many project teams are formed without any foresight as to the dynamics of individuals (e.g., experience with teams, learning and leadership style, communications, and work skills) or their current time commitments (e.g., personal interests and workload priorities)."

You must anticipate such problems. Therefore, any development schedule should assume that you will have to work with nonideal resources.

Your development work will often take longer than desired (or planned and expected) because the assigned personnel are less effective than desired.

> "The project required that several tests be performed on a number of product samples. A work order was authorized, and I then discussed the test plan with the technicians who would be running the tests in the analytical laboratory. I was assured that all the test work could be completed before my deadline.

(continued)

Your development project can be delayed by mistakes or inadvertently incorrect information.

The day before the deadline, I was told that the tests would have to be sent to a subcontractor since we no longer had the equipment required to perform this test. The result was that the development project was delayed."

Development managers may lack management training, skill, or interest, thus delaying a project.

"I have been trained in a technical discipline. Now I am expected to perform development work within my technical specialty, and I am also expected to manage other aspects of the new product development effort."

Conflicting priorities among development projects and a project that is staffed with people having other duties is, unfortunately, fairly common, so delay and a missed schedule are the normal result.

"My development project is largely staffed by people who also have full-time responsibility on other projects. Work on my project is handled only after it becomes a bigger crisis that the work on other projects. Everyone is therefore always working in a crisis mode."

These comments are illustrative of real problems. Although it is no panacea, project management tools and techniques can help overcome these and similar problems that await the unwary.

Continuous Learning

The third element shown In Figure 3-1 in the S&G interval is continuous learning. This can be applied usefully in the FFE interval (especially to discontinued efforts) and also in the PPS interval after the product has been on the market for several months. Because of the intensity of effort in the S&G interval, it is most appropriate at this time.

The objective of continuous learning is the retention of expensively gained knowledge. In Chapter 6 we treat continuous improvement reviews more extensively.

SUMMARY OF KEY POINTS

- The S&G interval is concerned with the management of a single project for which the target specification should be clear.
- The schedule for time-to-market depends on the difficulty of the specification and the availability and effectiveness of resources.
- Management pressures for changes and inadequate resources may delay the project schedule.
- The expected cost-of-goods to be sold is much more important than the project's development expenses.
- Project management software can assist with scheduling and resource coordination.
- Continuous improvement reviews normally should be conducted during the S&G interval (as well as at other times).

$$4$$

After Launch: The Preprofit and Continued Sales Intervals

INTRODUCTION

Once you launch your new product or service, you wish to achieve the profit objective quickly [a short preprofit (PPS) sales interval] and begin a long period where you can "milk the cow" or enjoy the "gravy" [a long continued sales (CS) interval] until the inevitable end of life. In this chapter we discuss both of these intervals, concentrating on the challenges and problems that can accelerate or delay achieving your profit goal. These problems include a production rate that is inadequate to meet customer demand, use of inappropriate distribution channels, and a lack of sufficient commitment by the sales force. Ways to shorten the PPS interval and stretch out the CS interval are described.

THE PREPROFIT SALES INTERVAL

The PPS interval starts with the shipment of salable products. It ends when you have achieved your firm's profit objective. The PPS period's goal is rapid achievement of the originally established profit goal.

Most NPD process descriptions show a pot of gold or a big dollar sign as the stage after a product starts routine product shipments. In fact, what has already been done in the FFE and S&G periods and what is still to happen in the PPS period are the critical determinants of how rapidly—if

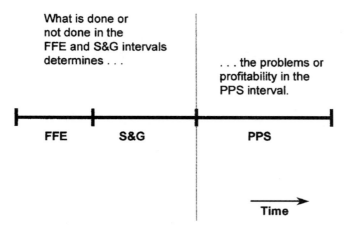

What is done or
not done in the
FFE and S&G intervals
determines . . .

. . . the problems or
profitability in the
PPS interval.

FFE   S&G   PPS

Time

Figure 4-1. *The PPS period is critically dependent on what precedes it.*

at all—that pot of gold will be reached. This is illustrated in Figure 4-1. Sadly, many companies disconnect the eventual sales activities from product development; and many more feel that the only linkage is the new product roll-out to the national sales force. A more intimate connection, built into the FFE and S&G activities, can increase sales, decrease costs, and shorten the PPS period.

Occasionally, a company will change the profit objective when the PPS period starts. "Oh well, we got 40 percent market share (versus the original target of 50 percent) and have the largest share." That may be a fine accomplishment, but it falls short of the target used initially to justify the effort. Sometimes this comes about because the multiple goals are somewhat ambiguous or partially contradictory. For instance, in this example, another profit objective goal might have been to have the largest market share, and the achievement of 40 percent share met that goal. The tendency is to reduce goals that are proving difficult to achieve, to "weasel out," and to declare victory when it has not, in fact, actually been achieved.

> *A myopic focus on user benefits during the FFE and S&G periods may shorten the time-to-profit during the PPS period.*

A product or service provides *deliverables* (e.g., performance, features, durability, serviceability, combined with timely availability and convenience) and its procurement or use involves *interactions* (e.g., courtesy, credibility, security, empathy, combined with responsiveness and accessibility). The price of these must be less than compa-

> *Do not lose sight of the original profit goal.*

rable alternatives. You can command a higher price only when the deliverables and interactions are better than ultimate users' perceived alternatives.

The profitability you achieve after the launch depends on quickly solving any customer or user problems. This requires a service orientation and a display of great courtesy when users make "dumb" mistakes using your product. You can help these customers and users by providing ease of access via toll-free telephone lines or equivalent mechanisms. Postlaunch profitability also depends on upgrading or enhancing the product in a timely way, with line extensions or other family members.

You must maximize real benefits and value to users and minimize your costs to shorten the PPS period.

The PPS period ends when the profit objective—whatever this (or these) may be—has been achieved. These is no sharp transition to different activities or demarcation in what you now do in the CS period, although the concerns and opportunities can change, as discussed later in this chapter.

PPS Interval Problems

The list of problems that can plague a new product when it reaches the market is almost endless. Software—especially for ubiquitous microcomputers—is frequently cited to provide examples. I mentioned errorware in Chapter 1, and another writer has stated: "Software may be the only business where companies routinely ship product with known defects" [1]. An executive in a telecommunications company is quoted as saying: "Rarely is software perfect the day it's rolled out" [2]. In fact, much software must be fault free when rolled out. Examples include almost all software on which lives are dependent, such as for flight controls in airplanes and controls for medical devices, including pacemakers. When introduced, software may lack embellishments that can be—or are intended to be—added later, but there is no fundamental reason why it must be inherently flawed. Mass market software, however, seems to presume that the initial users are willing to serve as guinea pigs. The then widely reported flaws guarantee that timid buyers will not voluntarily purchase the product and these flaws often encourage others to delay purchase. They think, "Why buy trouble?" So, continuing to introduce products that are known to be defective must delay the achievement of the profit goal.

Sadly, new types of problems (which may bring chuckles to amused observers) continue to emerge to frustrate prospective users. No list, therefore, can be exhaustive, but the following common problems are illustrative:

- Users encounter problems when they unpack or install the product, or needing setup assistance, find an unhelpful "help" line.

Here are two contrasting cases I have personally encountered to illustrate two extremes. To paraphrase Charles Dickens, I had the best of times and the worst of times attempting to use these products. American Airlines *AAccess* and Federal Express *FedEx* Ship are free software products that are apparently intended to provide both the users and suppliers with benefits. Although I have great admiration for and make use of both companies, my experience with these two offerings differed dramatically.

AAccess was announced and widely advertised (e.g., in *Business Week* and *Financial World*) in the summer of 1996. Disks containing the program were mailed to American Airlines' Platinum frequent flyers, the airline's best customers. *AAccess* was intended to be a flexible universal reservation system for individual use on a personal computer (PC).

The benefits of *AAccess* to American Airlines appear to include: the need for fewer reservation clerks; direct ticket sales that save travel agents' sales commissions, and the possibility of earning sales commissions for booking hotel and rental car reservations, which are linked into the system. My goals in testing the service were to save time and to gain access to an enlarged selection of flights, seats, hotel rooms, and rental cars.

Unfortunately, the initial version (1.0) could not recognize a Platinum frequent flyer. Consequently, flights that still had an inventory of unsold seats set aside for these premium flyers were identified as being sold out. One has to wonder why the initial launch plan targeted the best customers but lacked an essential feature to serve their needs. If this launch plan was considered in the FFE, it seems unlikely that this would have happened.

In early 1997 an upgrade was provided (version 1.1), but it was unable to provide connecting flights on American Airlines where another airline had direct flights, the help desk was staffed by a person (when I attempted to overcome this problem) whose English language skill was marginal and who failed to solve the problem, and finally, there was no evidence that Senior fare discounts (to which I am entitled) were accessible. If the plan to roll this product out to American Airlines' best (i.e., Platinum) customers was clearly identified in the FFE, it would not be necessary to, at best, inconvenience them. I have discontinued attempts to use this product.

(In mid-1998, American Airlines promoted the use of its Web site, www.aa.com, to book flights on American Airlines and American Eagle [3]. This service still had no way to recognize eligibility for a Senior fare, which required a separate telephone call. This inadequacy begs the

(continued)

question: If a user still has to make a call, why go to the trouble to use the Web site?)

FedEx Ship is a system that allows a person to use a PC to create a laser-jet-printed shipping label, store the recipient's information for subsequent use, and track the package. The benefits of *FedEx Ship* to Federal Express appear to include less data entry and therefore potentially fewer clerks, more shipping labels that are easy to read, and fewer misdirected shipments. My central goal in testing the service was to save time by storing frequently used recipient addresses. The increased assurance that the labels printed on a laser jet would be correct and easy to read was also attractive.

In dramatic contrast to the experience with *AAccess,* installation of *FedEx Ship* was simple and fast, and the help line (which I did need briefly) was calm, lucid, and immediately helpful. The product turns out to be very easy to use and I now make frequent use of it.

- Users encounter unanticipated difficulty while using the product.

The cup holder—when it was open—prevented the gearshift in the new 1998 Cadillac Seville STS from moving from second to low gear [4].

- There are unintended consequences.

Heart monitoring devices in a hospital blanked out the nurses' displays when a nearby television station started to broadcast high-definition TV signals [5].

A newly released Internet product that was intended to simplify Web address searches put users into a pornography site when they keyed in "Bambi" [6].

- The producer or, worse, the user is saddled with high maintenance costs.

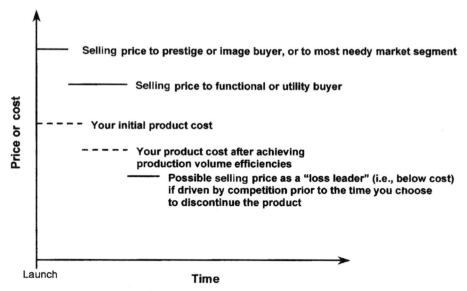

Figure 4-2. *The price you can charge changes in time.*

> The National Highway Traffic Safety Administration (NHTSA) required Chrysler in 1998 to recall some 90,000 1995 cars because NHTSA determined that seat belt fastenings were unsafe [7].

High maintenance costs will lengthen the time-to-profit.

- The producer or someone in the supply chain is forced to maintain a large inventory of spare parts.
- The producer incurs an unexpected product warranty cost and, perhaps, must cope with returned goods.
- The producer discovers that the price-to-cost ratio is less than planned. Figure 4-2 illustrates that the product cost normally will decrease over time. The price you can charge may also have to change in the presence of competition, and you may even find that it is necessary to sell at a loss at some point in the CS period.

If a product is easy to use, which will become apparent in the PPS period, you can gain word-of-mouth referrals. If it's fun to use, you may get even more benefit. If it's not easy to use, you incur extra, unnecessary customer service costs. Thus, if a product is easy to use, you get a double benefit: Increased sales and reduced costs.

So, why are some new products hard to use? Are product designers convinced that they can more easily fix problems later rather than during development? Presumably, ease of use is not seriously considered in the FFE and is thus not highlighted for sufficient attention during the S&G period.

Much of the user difficulty that is experienced in the PPS period is caused by not involving the customer support function during the FFE period and early part of the S&G period. Issues such as installation, user training, the quality and clarity of documentation and labels, periodic maintenance, and so on, do not get dealt with in many companies until very late in the S&G period [8]. Clear, friendly, and timely product support can be a critical advantage, especially for technology-based products for which the extent and nature of support is a consideration in the purchase decision.

Do some product-use difficulties inadvertently occur because "firefighting" is seen as urgent fun by developers?

The use of personal computers and their associated software provides a frequently encountered example of frustration with the often-required technical assistance [9]. In too many instances, the user documentation—whether available in a book provided with the product, in a "readme" file that accompanies the product, or accessible via the World Wide Web (for those who can locate it)—is wholly inadequate if not totally obscure. Access to knowledgeable technical help is therefore an important consideration for both vendors and users, and some of the considerations for both are shown in Table 4-1. If the use of the product is not intuitively obvious—and few are—the vendor must make decisions about how to provide assistance. Key questions are whether to provide the help and telephone connections for a fee or to cover these costs in the price of the product. This is the kind of decision that should have been made in the FFE period, since it may be critical to user satisfaction and affects the vendor's pricing, distribution, and both the size and training of staff.

Ease of setup and making any desired or required adjustments are other aspects that users must cope with after they have obtained the product. This is another example of issues that need early FFE or S&G attention to reduce needless PPS obstacles.

Unreachable connections and tiny adjustment knobs (or screws) are imposed on unwitting users by unthinking developers.

Other considerations may also be critical, such as the user's total lifetime costs and the likelihood of unscheduled downtime. The latter can be crucial for new products that are intended to be used in just-in-time (JIT) production operations (e.g., automobile assembly and production) or those that assess, examine, or measure short-lived samples (e.g., biological specimens).

TABLE 4-1. Considerations Affecting Vendor's Decision About Ways to Provide Technical Assistance

	Vendor		User	
	Pro	Con	Pro	Con
Technical advice phone				
Free 800 line	More likely to learn about user problems in a timely way Supports user-friendly image	Cost of 800 line Cost to staff the line Higher product price to cover costs of 800 line and technical support may drive some buyers (especially the most knowledgeable ones) to buy alternative products	No inhibition to call Attractive to less experienced users	Higher price for product than for products for which support is not "free"
Toll line				
High capacity		Cost of staff to handle large volume of simultaneous calls		Cost of toll call and idle time while waiting for technical support
Limited capacity	Less likely to be bothered by "stupid" calls			High cost of toll call and idle time while waiting for technical support
Technical advice				
Free		Cost to provide the help	Attractive to less experienced users	
Fee based	Source of revenue			Annoyance and cost to get help that should not have been required for a well-designed, well-documented product

A manufacturer of medical devices intended for use in doctors' offices was committed to selling its products in international markets. However, the control panel labels for a new product were entirely in English, as were the user instructions.

There are other problems in the PPS period—not due to product use per se—which are created in earlier periods. Examples are a mismatch of production capacity and sales volume, an inappropriate distribution channel, and the degree of commitment to the new product on the part of the sales force. These are discussed in the following sections.

Production Rate   It can be a challenge to match production rate to sales. If the product is very successful, it may be necessary to expand production volume rapidly just to meet demand or to keep competitors out of the market.

The 1996 launch of the Advanced Photo System cameras, a joint development effort by Kodak and other Japanese camera and film companies, was disappointing because there was inadequate production capacity by all the manufacturers to furnish the cameras. Given the massive advertising commitment for a simultaneous launch in Japan, the United States, and Western Europe, it was impractical to delay the launch. Initial enthusiasm for the promoted product could not be satisfied, and subsequent sales were disappointing. This situation was exacerbated by the high cost of specialized film-processing equipment that had to be installed by camera and film retail stores, because in many cases these retailers were unwilling to make the investment until prospective processing volume was assured [10]. In retrospect, more geographically localized product launches might have been less troublesome.

Merck's Crixivan protease inhibitor for treatment of AIDS was fantastically successful when approved by the Federal Drug Administration (FDA). The demand was huge—understandably—and treatment required that once a patient started to take the drug, he or she had to continue to take it. Therefore, Merck found it necessary to limit the enrollment of additional needy patients to assure an adequate supply to the initial patients. It required a major effort by Merck to add production capacity quickly and get this approved by the FDA to meet the rapidly growing demand.

Scaling up or expanding production too slowly can kill the market or hand sales to others. Conversely, building production capacity before the product is proven is risky.

An earlier book of mine sold out its first printing in a matter of a few months. Due to personnel changes at the publisher, four months then elapsed before a second printing was available. Given the generally short life of most professional books, this unavailability while demand was initially great killed future sales of the book [11].

A tiny company, Dauphin Technology, must invest in production capacity before the performance of its initial units are proven so that it can meet buyers' conditions to provide large volumes if the trials are satisfactory [12].

Baxter International reportedly invested over $100 million in a production facility to manufacture a blood substitute that was in development [13]. Unfortunately, the new product was ineffective, so the new facility is presently idle.

The sales forecasts for the new product may have been overly optimistic. In some cases, users are locked in to alternatives because of switching costs, even if the alternatives are less useful.

My office continued to use WordStar for many years after Word (and WordPerfect) were the more dominant software packages for word processing. It was less costly to me—in terms of my time—to train new clerical help on WordStar, which I had mastered, than for me to learn how to use Word.

Distribution   The broad choice you have for distribution of your new product is whether to launch it into new or existing channels. Channels include manufacturers' representatives, stocking distributors, retail stores, mail order, your own sales force, and now Internet sites (such as Ama-

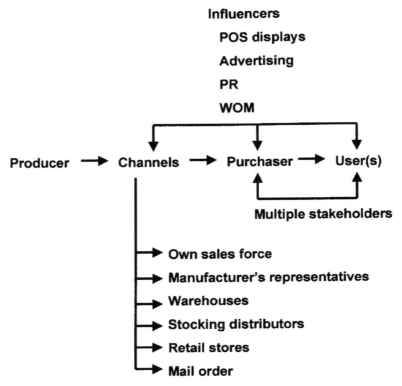

Figure 4-3. *Some distribution channels and purchase influences.*

zon.com). Figure 4-3 illustrates some of these channel options and factors that influence the purchase decision. There are subsidiary issues that must also be considered and—hopefully—resolved in the FFE or initial S&G period activities. You must decide where the product should be distributed initially (e.g., the United States or outside the United States). (In the earlier case of American Airlines, their initial target buyer— the Platinum flyer—was inappropriate for the product's initial characteristics.) Similarly, you must know when and how it should be promoted (e.g., branding, extent and type of a public relations effort, point-of-sales displays, and advertising content, frequency and location), and what price should be charged initially.

> *The distribution channels you intend to use affect the design and documentation required for the product.*

Researchers studying the launch of new physical goods in industrial markets in the United Kingdom have found that there are implications for the initial pricing strategy and distribution. If the product is more innovative than the competition, a skimming pricing strategy is best. However, if the product is not distinctively more innovative than the competition, a penetration strategy is to be preferred. Not surprisingly, they also found that existing channels may be best for products that have incremental improvements; whereas new channels may be better for a product that departs significantly from past and current products. In short, many distribution considerations—including explicitly who is responsible for what—all contribute importantly to how long it takes after market entry to achieve the profit goal [14].

Another researcher, examining consumer goods, divides these into two categories: functional, for which he asserts that demand is predictable, and innovative, for which demand is unpredictable [15]. The former category includes consumer staples, while the latter includes premium products and things like personal computers. In terms of distribution, functional products require a physically efficient process with a very low distribution cost. Innovative products require a distribution process that is rapidly responsive to the market's unpredictable demand. The wrong choice of the distribution channel imposes costs on the producer, thus delaying the achievement of the profit objective.

<table>
<tr>
<td>Distribution channels may change after a product is established in the market. As an example, professional camera film is initially targeted at a small population of heavy users with demanding requirements [16]. Later, the same film may be attractive to the advanced amateur and then, ultimately, to the larger consumer market of occasional picture takers, at which point promotion, advertising, and distribution may all have to be adjusted.</td>
<td>*The distribution channel must be appropriate for the product and its market.*</td>
</tr>
</table>

Sales Force Commitment   A highly motivated, knowledgeable sales force can obviously help build sales volume rapidly. A research study has found that a salesperson's commitment to selling a new product is enhanced when they have a stake in the new product, by age, and by experience. Their commitment is diminished if the product is highly innovative (since, presumably, they would have to invest a lot of time to explain the product's benefits) or if there is ambiguity about the customer or the buying process [17].

An experiential report suggests, in fact, that innovative products are plagued by an inappropriate announcement to the sales force. Typically, this

The product launch must stress users' advantages.

announcement stresses the novel and exciting features of the new product. Instead, the emphasis should be placed on the needs of customers and users and on how the product can satisfy those needs and solve problems [18]. New technology and other "jazzy" features do not benefit users unless these help them do a better job or produce a better product.

How to Shorten the PPS Period

A way to shorten time-to-profit is to make the product or service fun to use and simple.

> Many years ago (the mid-1960s if I recall correctly), Polaroid produced a camera that had four numbered knobs and levers. All you had to do to use it was to adjust these in sequence, and the last (i.e., number 4) was the shutter lever. This was as close to "idiot-proof" as possible and produced fine instant photos every time.

Invest time to make it obvious how to use your new product.

> An aircraft weighing kit that I had to use at about the same time provides another example of an easy-to-use product. It came in a metal carrying case with a clear picture inside the cover showing how to connect the various parts. Each connection was color coded, and the only technical obscurity was that the user had to know the meaning of a "tare weight."

Other products illustrate situations in which users can be frustrated or the developers can become disconnected from the users.

> The Internet provides—in my view—an example of a product that is deeply flawed today. Its use is frustratingly time consuming, not intuitively obvious to nontechnically adept people, and can be costly in those cases where the user pays for connect time by the hour.

Consider the purchase of a car. Is this a purchase process that provides fun? Executives at General Motors used to get cars delivered to them automatically. (Perhaps they still do, but I don't know.) Consequently, these executives never experience the normal buying environment. In addition, the company, precluding experience with that common, and often unpleasant, user experience, automatically services their cars.

Finally, there are actions that can and should be taken in the FFE and S&G periods to avoid PPS problems and thus shorten the time-to-profit. Again, the list could be almost endless, so a few illustrations must suffice:

- Design the product for ease of use, which can—and should—be established by testing early prototypes with a variety of users.
- Aim for a product design that reduces variation, is easier to manufacture, and therefore has less warranty expense.
- Spend enough time to test and assure that the design is truly both consistent with the requirements and that the requirements are correct.
- Three-dimensional design can be employed effectively to detect and remove a pending mechanical interference before any parts are fabricated.
- Invest in really excellent user manuals that are clear, well illustrated, and well indexed.
- Maintain a toll-free telephone help line that is well staffed to provide real help and not make callers feel foolish or inadequate by defending the company or belittling the inquiry.
- Make follow-up telephone calls or do a mail survey to measure the performance of help line personnel by the quality and effectiveness of the help that is rendered, not by the quantity of calls that are disposed of quickly.

In late 1998, after I reluctantly upgraded my computer's office suite, I encountered a problem that had not been present before. Suddenly I was not able to print large files that printed correctly with the previous version of the office suite. The help line spent a lot of time (for which I paid the telephone connection fee) but was not able to resolve the problem. I devised a somewhat time-consuming but usable "work around." A few days later, the company called to get my assessment of the help I received,

(continued)

and I expressed my unhappiness. Shortly afterward I received a call from, presumably, another help desk person, who was determined (or directed) to fix the problem. Unfortunately, when all was said and done, he chose to blame my hardware and file size—which, of course, was not a problem before installing the upgrade of their product.

· Plan for rapid product distribution after launch (or selective limited distribution with timely expansion if that's the preferred plan).
· Facilitate the efforts of the sales force by providing helpful selling tools and training, conducting an active promotion and advertising campaign, and otherwise stimulating their motivation.

THE CONTINUED SALES INTERVAL

The CS period starts when you have achieved the new product's profit objective. It ends when the product is withdrawn from the market and you discontinue all support functions for it. This period is—so far—rarely and sketchily examined in the NPD literature. A leading text [19] discusses "postlaunch control" but states that this activity ends when the new product has achieved its objectives. There may be other NPD books that provide some coverage of the CS period, although I have not yet found them. Perhaps this period is considered by others to be "just conventional marketing" and not part of the new product development process. Although the marketing function may indeed be heavily involved, other functions must also participate. The CS interval does, however, provide an opportunity for new product sales and profits—and the enhancement or lessening of your firm's reputation. Figure 4-4 sketches some broad options for the CS period.

Some companies are not very proactive in exploiting the CS period, where it is viewed as unglamorous or "ho-hum." Product management may be turned over to less creative or relatively inexperienced personnel. Alternatively, opportunities for the CS period may initially have been recognized in the FFE period. If so, these are normally first capitalized upon in the PPS period and exploited further in the CS period.

Once the profit objective is achieved, further sales can provide profit "gravy."

Conversely, some products are really designed to exploit the CS period. The proverbial razor blades that capitalize on the initial development of a razor [20] provide one example. New film formats depend on the initial development of a camera. Laser and ink jet printers generate a huge demand

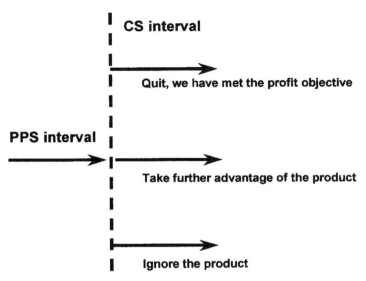

Figure 4-4. *Generic options for product during CS period.*

for toner and ink cartridges, and the market competition in this office supplies category suggests that it is extremely profitable [21].

Many food products are dated, encouraging use of purchased products before the expiration date. Although this is presumably done primarily to reduce the likelihood that spoiled food will be consumed, it may also stimulate sales and use. Other products (e.g., razor blades for Gillette's Mach 3 razor and some toothbrush bristles) change color at the time the manufacturer suggests they should be replaced [22].

The label for a pest-killing powder (Amdro for eradication of the fire ants that are prevalent in the southern United States) states that it remains active for only four months after the vapor seal is broken. Batteries for flashlights, calculators, and similar devices carry an expiration date, as do many over-the-counter drugs (e.g., aspirin, cough medicines) and prescription medicines. The efficacy of these kinds of products degrades with time, so dating promotes early use and more consumption. Photographic film also carries a "use before" date, although storage in a refrigerator can extend its useful life substantially.

The basic goal of the CS period should be to make money, not just get stuck in a rut and perpetuate the sale of the product. You can cannibalize the product if you wish, since the profit objective has been achieved. Or, you can use its existence and profit contribution as leverage to create other streams of revenue. For example, you can provide a new product as a replacement for an earlier model, provide a sales credit for the replaced earlier model to reduce the buyer's net purchase price, then refurbish the earlier model and offer it for sale to others.

> Many years ago, the Perkin-Elmer Corporation would accept trade-ins of an earlier model of their Micralign photolithography projection printers (used in the production of microcircuit chips during the semiconductor fabrication process) for a purchase credit toward a later model that could handle larger silicon wafers.

The CS period also provides an excellent opportunity to continue to sell required consumable products that are required to operate the original product. Obviously, these can themselves be considered freestanding new products, but they are more usefully viewed as desirable companions to or an integral part of the basic new product. Obvious ubiquitous examples are razor blades (for razors), photographic film (for cameras), refills (for mechanical pens and pencils), and ink or toner cartridges (for computer printers). You can also offer service and maintenance contracts. If this business is attractive, you want the CS period to be as long as possible.

> In early 1998, Hewlett-Packard branded MultiPurpose paper for ink jet, laser, fax, and copiers was offered for sale and promoted with introductory discounts. This additional Hewlett-Packard product presumably is manufactured entirely by a paper company (or companies) under contract to Hewlett-Packard, and may even be distributed by a paper company to the wholesale and retail outlets.

The CS period provides an opportunity to sell other products required to support the original product.

Not every product with a long life has a profitable CS period. Automobile tires, which have increasingly long lives, reduce the market for replacement tires [23]. The cost to produce a new generation of dynamic random-access memory (DRAM) chips may easily be $1 billion or more, but due to intense market competition, essentially all profits must be earned in the first year of sales [24].

Positive consumption externalities also can provide opportunities [25]. The more common and widespread your new product is, the more extensive a service and support network can—and must—be. The increasing prevalence of notebook computers among airline business passengers is leading to the provision of electrical outlets in the first- and business-class sections of aircraft, where these users are most likely to be seated. The need for but high cost of fast data transmission diminishes the Internet's attractiveness but will presumably lead in time to lower-cost fast data rate transmission [26]. This cuts two ways, of course. A new automobile (e.g., the DeLorean several years ago) may lack a network of dealers able to provide spare parts or trained service technicians, thus inhibiting sales. An initial trial of smart cards (plastic cards similar to credit cards but with cash electronically embedded in the card's magnetic stripe) in New York City was not successful, as neither merchants nor users found any compelling advantages of smart cards over cash or credit cards [27].

Eventually, a product will reach the end of its life. Clearly, a firm should not just drop all support for that product without providing some kind of timely notice.

In May 1998, Kodak announced that it would stop making film for the Instamatic 110 format cameras, effective at the end of 1999 [28]. These 110 cameras had been on the market for 35 years. At the time of Kodak's discontinuation announcement, Kodak provided any remaining 110 users with advance notice of a year and a half.

In June 1998, the Official Airline Guide (OAG Worldwide) announced that it was going to discontinue its electronic version. This service was accessible directly with a modem and via at least one Internet gateway (CompuServe), and it permitted a user to locate airline flights and hotels quickly and economically anywhere in the world. I used this very convenient and feature-laden product whenever the need arose, although episodically, after discontinuing my subscription to the domestic and international paper versions, which I had used for many years. The reason given by OAG Worldwide was that the electronic version was not year 2000 compliant. In a letter dated June 1, they supposedly enclosed a sample CD-ROM version that was year 2000 compliant and would be available by subscription with monthly updates. The CD-ROM sample was not enclosed (nor did it arrive separately), and a subsequent letter dated June

---

(continued)

> 5 reiterated the reason the product would be discontinued but did not mention a sample CD-ROM. In a subsequent query to their Internet Web page, I discovered that a free trial sample CD-ROM could be obtained by calling an 800-number, which I did. The overall impression, of course, is that OAG Worldwide did not carefully plan the discontinuance of its electronic version.

Until the end of a product's life, there may be legal requirements to provide spare parts and furnish some kinds of customer service [29]. If there are no legal requirements, there still may be sound business reasons (e.g., goodwill for the future) to provide product support or assure that it is otherwise available. Customer happiness can promote loyalty and future purchases of additional new products and support services [30].

SUMMARY OF KEY POINTS

- Things that are done or not done during the FFE and S&G intervals can create postlaunch problems that increase your costs and extend the PPS interval.
- Ideally, your new product's production rate will match market demand; otherwise, you will have excess capacity or there will be unfilled market demand that may be amenable to competitors' alternative offerings.
- The choice of distribution channel(s) should be made in the FFE interval or the earliest portion of the S&G interval.
- Have a plan to obtain sales force commitment well before launch.
- Try to devise a product or service that may extend the CS interval.

Part 3

Improving Your Process

5

Implementation

INTRODUCTION

This chapter covers general topics that bear on your ability to shorten the time-to-profit and exploit opportunities discussed previously. Some ideas are new or, I believe, helpfully clarified. Some ideas are reiterations of fairly well-known practices used in the best companies.

Surprisingly, many of these proven techniques are not yet used in many companies that would benefit. Perhaps this is because many executives and managers do not know of the advances that have occurred. Perhaps it is because innovation of new products is imagined to be inherently creative, and creativity is viewed as inherently undisciplined and disorderly. In fact, the new product development field is moving away from "art" to much more effective, systematic approaches, as depicted in Figure 5-1.

Successful new product development (NPD) requires many things, such as a feasible new product idea aimed at a market that will be receptive; a team that can design, develop, produce, and deliver the product; and a process that is not ad hoc to assure that this can be done quickly and repeatedly. Obviously, you can have all this in place and be unsuccessful, although making some inappropriate decisions at some point normally causes failure. And you can occasionally achieve market success with a marginal idea, developed by a comparatively ineffective team, using a totally idiosyncratic, random process. But the odds of success favor the company that takes advantage of proven NPD processes, tools, and techniques.

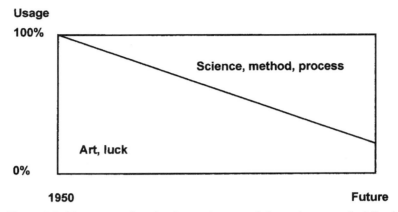

Usage

100%

Science, method, process

Art, luck

0%

1950   **Future**

Figure 5-1. The new product development process is becoming more disciplined.

RESOURCE OVERLOADING

More projects are often initiated in the mistaken belief that these will provide more chances to succeed or hit a winner; in reality, it means more chances to fail because they are all staffed with or have access to insufficient resources. The key to assuring a short time-to-profit is the way a company's inherently limited resources are allocated. Both resource flexibility and capacity are limited, as illustrated in Figures 5-2 and 5-3. In the case of baseball, we can ask how many great hitters make great pitchers or how many catchers make great shortstops. In the world of new product development, not many highly trained specialists are flexible generalists, so an available mechanical engineer may not have useful skills for a particular development project. Or, a critically required test facility may not have the capacity to accommodate two development projects simultaneously.

Are your NPD processes and systems designed to prevent NPD teams from making mistakes or to help them quickly succeed?

Resource allocation is the critical executive challenge.

Individual resources—physical and human, but principally the latter—are easily overloaded, as illustrated in Figure 5-4. Sometimes this is accidental, but sometimes it is a deliberate management tactic (which I find abusive). Figure 5-5 illustrates the requirement for resources (shown as A, B, and C, whatever these may be) to work on a single NPD effort during some period of time (e.g., a month or calendar quarter). When there is only one project in a company (e.g., a venture startup) it is relatively easy to anticipate an impending resource overload, as is shown for resource C (which could be either human

- **How many great hitters make great pitchers?**
- **How many catchers make great shortstops?**
- **Not many highly trained specialists are flexible generalists, so an available mechanical engineer may not have useful skills for a particular NPD project.**

Figure 5-2. Resource flexibility is limited.

Test facility
(e.g., test fixture)

Project

Big projects require more time in a constrained test facility

Big
project

Multiple projects (e.g., from different SBUs) require sequencing

Project A

Project B

Project C

How do you decide which
project gets priority?

Figure 5-3. Resource capacity is limited.

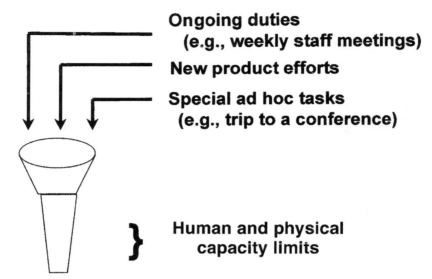

Figure 5-4. *Resource overloading.*

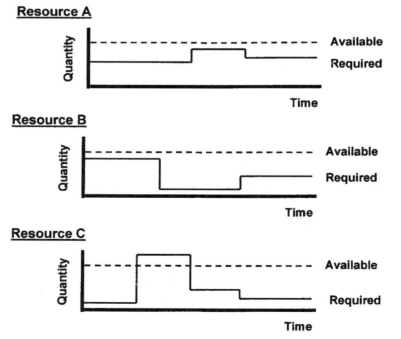

Figure 5-5. *Resource use versus time for a hypothetical NPD effort.*

or physical) in Figure 5-5. This may also be the case when there is only one NPD effort in a business unit of a large company (e.g., in a division). However, when there are several NPD efforts running concurrently, it is much harder to get visibility of the total demands on individual and collective resources. The result is that resources become overloaded, and when that happens, NPD projects are delayed.

In the next sections we elaborate on the resource allocation problem, suggest some possible solutions, and indicate considerations in establishing a resource allocation strategy.

THE RESOURCE ALLOCATION PROBLEM

In most companies there are too many development efforts under way at any given time, so all are starved for resources and move forward at less than the fastest speed possible. The challenge is to start only as many projects as can be completed in a timely way, as suggested in Figure 5-6. This requires executive discipline to defer unimportant projects, to suspend work on less important projects, and to kill some projects, which

Concentrate limited resources on a few projects, bring these to completion quickly, and move on to whatever projects are next most critical.

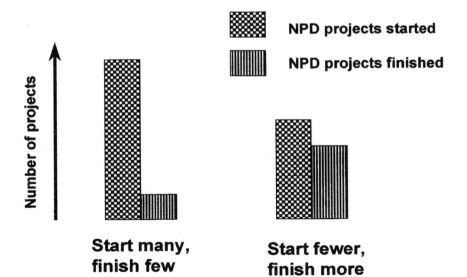

NPD projects started

NPD projects finished

Number of projects

**Start many,
finish few**

**Start fewer,
finish more**

Figure 5-6. *When your firm concentrates its limited resources on fewer projects, it can complete more of them quickly.*

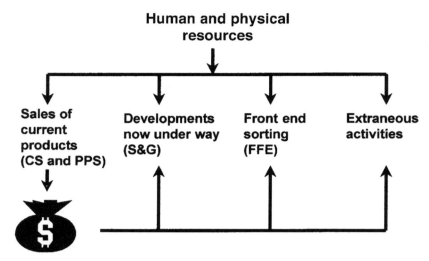

Figure 5-7. *Senior executives must make the choices about how to allocate the firm's resources.*

may include some development efforts that have been under way for a long time. Lazarus projects—those that continue to rise from the dead—must be killed.

In addition—perhaps worse yet—any problem with an existing product threatens immediate revenue and becomes a current crisis. This normally preempts development resources and thus interrupts some (or all) new product development efforts. Not uncommonly, longer-range development projects are sacrificed for shorter range. Figure 5-7 illustrates the key executive challenge, the allocation of limited human and physical resources.

The revenue stream that funds all else comes from products in their PPS and CS periods. (This ignores investment income, which can be significant in some diversified firms.) Extraneous activities that consume money are present in every organization, and include such things as vacations, holidays, sickness, and staff meetings other than those related to product development. What's left over must be divided in some fashion between the S&G and FFE periods and—within those periods—allocated to selected development projects.

Three Solutions to Consider

Let's briefly review the use of project priorities, project management software, and relationship maps, which are three means to at least reduce management's propensity to make unrealistic commitments of limited resources.

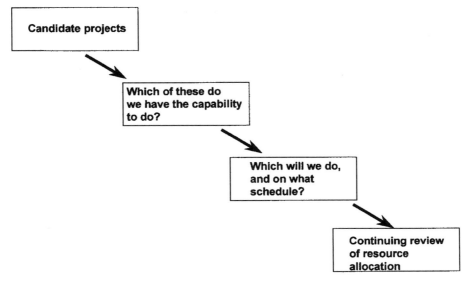

Figure 5-8. *Decision sequence to set project priorities.*

Establish Project Priorities   The benefit of project priorities that are arrived at thoughtfully is that development resources can be applied first to the highest-priority effort. If the highest-priority project never has to wait for resources that can be utilized productively, it will be completed faster than otherwise. Thus, the resulting product can begin to earn income, and some resources dedicated to it can now be redeployed. These resources are then applied to the second priority, and so on. (Obviously, if a company has enough resources to support several development efforts simultaneously, it should do so, but the practice of concentrating inherently limited resources on a smaller number of efforts so that these move faster still provides an overwhelming advantage.)

A new product development manager in one company stated, "Priorities change faster than work can be completed." This situation is not confined to that manager or even that particular company. The result, of course, is individual frustration and less effective corporate performance.

Figure 5-8 depicts a decision sequence that can help you accomplish this. Although the management actions and choices may be obscure today, there are corporate examples of the benefit obtained by setting a clear project priority.

In the late 1960s and early 1970s, Perkin-Elmer Corporation's divisions that performed development work under contract to various customers, primarily government agencies, were able to grow faster than the purely commercial divisions. A corporate strategy at that time was for the contract and commercial businesses to remain approximately equal, so that each of the contract divisions was directed to identify a new commercial product (based on the division's technology) and develop it as a business. The selection process in the divisions differed, but each had a few months to select the opportunity and initiate development. Thus, each division had a single commercial NPD effort that was clearly designated, that was sufficiently important to be reviewed for progress quarterly by the corporation's executives, and that was judged to be a major impact program. The outcome was one huge success, two so-so efforts, and two that were essentially written off.

Priorities must be set by corporate executives and general managers to preclude parochial influences by single departments.

In the late 1980s, 3M's Industrial Specialties Division employed a priority system for NPD projects that did not require an invention. One project in each major business segment could be designated its top-priority project, and all documents got a FASTRACK sticker attached to assure prompt access to required support [1]. Although I do not know the details of how the NPD projects were selected and the outcome, this seems to have spawned the current 3M Pacing Plus system [2]. The corporatewide goal of these accelerated projects is to produce significant changes in competition in markets for products with major sales and profit potential by giving them priority for use of 3M resources.

Figure 5-9a provides a grid in which all current projects can be arrayed, and Figure 5-9b is an illustration of one way in which it might be employed. No projects appear in the PPS or CS columns in Figure 5-9b because none are planned (until the need is imposed by some problem), but other second parameters (e.g., newness to market or the nature of the competition) would have entries in these columns. A company might use several of these arrays, each with a different second parameter on the vertical axis. Each project can be shown as a dot, or—more helpfully—as a circle, the size of which is proportional to its payoff, required investment, or some other significant

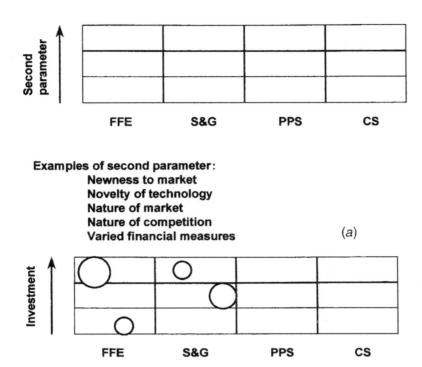

Examples of second parameter:
Newness to market
Novelty of technology
Nature of market
Nature of competition
Varied financial measures

(a)

Each circle represents a project.
Size of circle is proportional to anticipated after-tax income.
The small return expected for the large investment project
in the S&G interval should be questioned.

(b)

Figure 5-9. (a) *Grid in which project parameters may be displayed visibly to clarify the distribution of resources among existing and pending development projects; (b) example of the use of the grid.*

characteristic. Candidate projects, those not yet authorized, can be shown (presumably in the FFE period column) with some other symbol, if desired. These candidate projects might be identified by the use of product road maps or other future scenario planning tools.

Project Management Software   Project management software is one of many project management tools that has great utility in speeding new product development. Every new product development effort is a project. Project management software is most useful in the S&G period. It is also

Do not let pet projects preempt important projects.

helpful in assuring that any series of tasks in any period are defined and planned before the work is undertaken. It may be least useful for new-to-the-world product development efforts, where it is never very clear what has to be done next. For many new-to-the-world projects, every subsequent activity may depend on the unpredictable outcome of some trial, test, or exploration. However, even in this situation, project management software can be used to plan for the work that you intend to perform during these probes [3].

Resource coordination was discussed in Chapter 3, and Figure 3-11 illustrated one use of project management software. As pointed out there, project management software can be used to encourage multifunctional teamwork by serving as both a coordination and responsibility matrix. When used this way, project management software can be used to identify resource conflicts before they occur. When the resources required to carry out the planned work are entered, the software will provide a forecast for the required amount of any resource and when this resource is scheduled. Thus, project management software can provide data such as those shown in Figure 5-5 and it can do this for a single project or for a group of projects. A resource requirement (e.g., software engineering or environmental test chamber capacity) that exceeds what is available at a particular time is a clear signal that the development plan will not work successfully. Ideally, this resource forecast will be done while there may still be time to devise an alternative plan that avoids or minimizes delay. Figure 5-10 illustrates an approach to clarifying resource adequacy (or overload), which can be constructed for each key skill or unique facility, or at a macro level for the sum of all corporate resources. The right-hand side shows available capacity, which is clearly less than the estimated resources required for additional opportunities shown on the left-hand side of the figure. In this situation either some opportunities must be defered or external resources must be obtained (as shown in the upper right of the figure). How resource loading and overloading is actually displayed depends on the particular project management software that is used. In the case of Microsoft Project, which is widely used today, resource requirements are shown as bar graphs for each time period selected, and overloads can be clearly shown in red.

Team communication can easily become critical and is often time consuming when external resources are part of the multifunctional project team. (These may be called virtual teams in some situations.) Electronically linked data files and frequent e-mail or other forms or telecommunications can helpfully facilitate the work of a geographically dispersed team, but these aids are no substitute for face-to-face meetings. Thus, you must schedule periodic meetings of all key personnel and include both the cost of and time

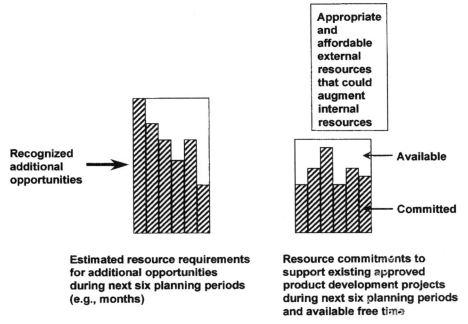

Figure 5-10. *Resources required to support new candidate projects can overload the available resources and delay some or all new product development projects unless the timing of resource requirements is examined in detail.*

for these in the NPD project plan. It has been my experience that these cost and time investments will save far greater costs and delays caused by "penny-wise, pound-foolish" economies.

Relationship Maps Relationship maps can be used effectively to clarify organizational roles in the development of new products. The idea for relationship maps, which have also been called process maps, is not original [4]. However, the idea may have great value for new product developers. Figure 5-11 shows how one might look. Starting at the top right, you see that the marketing function is charged with making proactive inquiries among prospects and users about their needs and wants, their knowledge of the competition, and any trends of which they are aware. Similarly, prospects and users may themselves initiate discussions with either the marketing or the research, development, and engineering (R,D,&E) functions about their needs and wants, either current or future. The rest of the map indicates other information and product flows.

Figure 5-11 is only illustrative and does not show many other functions that normally play some role—often very important—in new product devel-

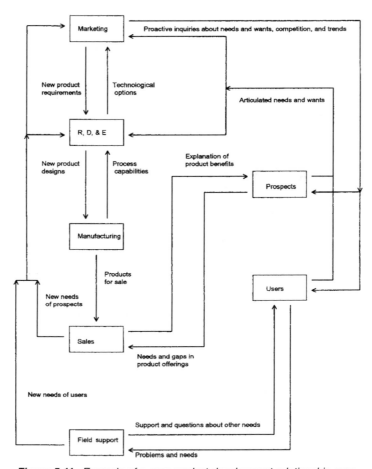

Figure 5-11. Example of a new product development relationship map.

opment, such as human relations, finance, or quality. These have been omitted for simplicity; in each company, these functions must be added as appropriate. For example, human relations plays a role by delivering qualified employees and providing ongoing training. Also omitted are many external influences that should be considered and, perhaps, included. As one example, external technology (i.e., emerging, novel for firm, or similar to what is already employed) is normally brought into the R,D,&E function.

There is no distinction among the various kinds of prospects in Figure 5-11. A more complete map might have to distinguish among current users, current nonusers, and those who have become ex-users. Nevertheless, the relationship map clarifies functional roles and may help in the design of a new product development process.

RESOURCE ALLOCATION STRATEGY

Every company, even the very largest, has limited human and physical resources. Some of these resources must be devoted to managing the existing products and markets, and dealing with the day-to-day distractions that consume time (e.g., staff meetings, holiday and other social events, personnel issues). Resources that are left over are available for new product development. In most companies, however, these resources are not allocated thoughtfully among:

- Minor product refinements, substantially new products, and major breakthrough efforts
- The FFE, S&G, PPS, and CS periods
- High- and low-risk efforts
- Near- and long-term projects

The growing literature on portfolio management explores this in more detail, but there is no magic formula [5]. Resource allocation is *the issue* that top executives must actively manage: The use of project management software can elucidate specific potential resource shortages; thereafter, specific project priorities must be assigned to help resolve or overcome these shortages.

Opportunities arise asynchronously from the timing of annual corporate planning. It is therefore helpful to have available capacity to begin work on unplanned attractive opportunities when these arise, but this is a costly luxury that is not normally tolerated because it is seen as a waste of resources. Alternatively, a system that allows for the reallocation of resources in a timely fashion can be employed.

> If there are no surplus resources to handle additional new product development projects—which is the normal situation—personnel are frustrated and work is delayed, as the following comment from a practitioner illustrates: "Constant interruptions for 'new' priorities delays results but not expectations."

Resources—human and physical—do not have infinite capacity to handle more work.

MULTIFUNCTIONAL TEAMS

The sine qua non of new product development is the effective performance of multifunctional teams. The team may be large or small, depending on the proj-

• Small or simple

- – May require generalists
- – Less experienced PM may be adequate

• Large

- – Higher intrateam communication "overhead" cost
- – Investment to promote teamwork is required
- – More experienced PM is required

Figure 5-12. *Team size issues.*

ect's complexity and schedule. Figure 5-12 indicates some of the issues that you should anticipate, depending on the size of a multifunctional team. The experience level of the team leader or project manager (PM) is generally the critical factor. Figure 5-13 illustrates another aspect of this. Note that no absolute numbers are used in these two figures, since it is somewhat relative. However, six or fewer team members would normally qualify as a small team, and more than two dozen would almost always be seen as a large team.

Team size and other characteristics also depend on the project's duration. Figure 5-14 summarizes a few project duration issues related to multifunctional teams. The central challenge in many cases is with the timely availability and motivation of people who are assigned to the new product development project while simultaneously having other duties. Such personnel often have conflicting priorities. This reality is what leads many consultants and prac-

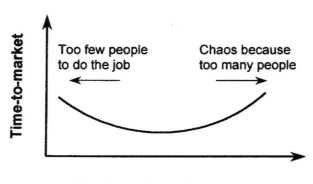

Number of people on team

Figure 5-13. *Time-to-market depends on the size of the team.*

- # Short
  - **Frequently dependent on part-time people**
  - **Hard to justify team collocation**
  - **Normally not practical to change PM during project**

- # Long
  - **Can justify full-time, dedicated people**
  - **Collocation is possible**
  - **Often change PM as work progresses**
  - **Risk of boredom or burnout**

Figure 5-14. *Project duration issues.*

titioners to counsel that personnel should be dedicated to a single project. Unfortunately, this is not practical in many companies, because small projects that cannot justify the use of personnel on a full-time basis are very common.

A practitioner's observation and concern about problems that he recently experienced underscores another aspect of this issue. "The majority of projects we are working on are related to achieving manufacturing cost reductions. While the primary projects are assigned to specific individuals, most projects require inputs from many different plant personnel and also corporate staff (at other locations). How do we get people to participate on projects that are not specifically assigned to them?" The answer to this kind of question is motivation. The NPD PM must try to identify some facet of the project that will be of interest [6].

Figure 5-15 shows some considerations about team size and project duration that depend on one another. The frequently cited new product development timeliness panacea of collocation is not, in fact, suitable for short-duration projects. Collocation can be used for a small team on a long-duration project, but it may be more helpful to expand the team's size. When a team includes members from dispersed sites with specialized facilities upon which they are dependent (e.g., scientific laboratory apparatus, model-making machinery, production equipment), it may be

Collocation can be helpful, but it has risks.

		Short	Long
Team Size	**Large**	Often a hasty reaction to marketplace changes Risk of chaos and wasted money	Team collocation is often required and justified
	Small	Normally is low cost	Must determine if a larger team could shorten the schedule or divide into subteams for specific deliverables
		Short	**Long**
		Project Duration	

Figure 5-15. Team size and project duration issues that interact.

impossible to collect all people at one location. A collocated team working at a site remote from corporate services it requires will often be impeded.

One solution to achieving effective performance from a multifunctional team that is dispersed is the use of various electronic linkages. Many companies have the capability to share critical design data files and other essential product development information at multiple sites, including international locations.

Ford Motor Company provides one example of a company that links designers at remote locations. Workers in Turin, Italy and Dearborn, Michigan can work interactively on designs before more expensive models must be constructed, thus saving both time and money [7].

The use of multifunctional teams can produce a challenge for many executives and managers: rewards to promote cooperative multifunctional teamwork. The challenge is especially great in large established corporations, where much management attention and energy is devoted to running the ongoing business. Frequently, that business is functionally organized, the corporate culture stems from this organization, and an employee's career succeeds or fails in part by his or her functional loyalty. Teamwork may be less of an issue for a small startup company, where the sole critical activity is getting an initial product to market after which stock options can provide very large financial rewards upon market success.

While corporations continue to stress multifunctional teamwork, they continue to reward primarily individual behavior. Salary and promotion are normally decided by functional department supervision, and it is often highly idiosyncratic. Recent preliminary research [8] suggests more effective practices:

- Compensate top management, in part, for new product development success.
- Employ a combination of intrinsic (nonmonetary) and extrinsic (monetary and positional) rewards.
- Reward teamwork if you want to promote it.

Several years ago, a midlevel manager was chosen to lead an important multifunctional new product development team in a large corporation. The effort was ultimately unsuccessful, primarily because the strategic choices made by the corporation's senior executives were flawed. Instead of recognizing the team leader's increased knowledge and experience after this effort, his management left him in the corporate "doghouse" for a few years. Many other team members left the corporation. While the team leader's career subsequently revived, the doghouse message did not encourage—and noticeably discouraged—risk-taking new product development efforts in that corporation.

What you provide "strokes" for is what you get more of.

NPD project leaders of some multifunctional teams will experience difficulty because they lack supervisory formal authority. This problem may be due to their perception rather than based in reality, because formal supervisory authority is ineffectual (if not counterproductive) in many cases. Nevertheless, project managers in these situations have asked such questions as: "What recourse does a project leader have when a team member is reluctant to conclude a project-related activity in a timely manner? and "If project leaders are to be held accountable for the project completion, should they not be allowed input to the annual performance reviews of their team members?" The key to answering these and similar questions, and resolving the underlying problems, is for the project leader and his or her management to foster a team spirit. Off-site social events (e.g., picnics) illustrate one technique.

TOOL APPLICATION

There are many tools that can be used to shorten time-to-market and time-to-profit. None is sufficient of itself or works in all situations, some may not work at all in your organization or on your specific NPD project, and you may even be successful without the use of any of these. However, each has proven valuable to some professionals in at least some NPD situations, so you should understand enough about them to decide if they can help you. These tools include:

- Strategic planning
- Multifunctional teamwork
- Supplier involvement
- Customer and user involvement
- Marketing research (e.g., focus groups, conjoint analysis, choice models)
- Quality function deployment (QFD)
- Specification setting (both qualitative and quantitative)
- Product families
- Computer-aided engineering, design, and manufacturing
- Failure modes and effects analysis (FMEA)
- Design of experiments (DOE)
- Design for manufacturing and assembly (DFMA)
- Robust design
- Zero defects
- Life-cycle design
- Total quality management (TQM)
- Product integrity
- Test plan and testing
- Documented new product development (NPD) process
- Project priorities
- Resource capacity planning
- Project management software
- Discounted cash flow financial analysis (IRR, NPV, or similar)
- Activity (ABC) costing
- Target costing
- Risk assessment

Some of these tools are most useful in specific periods rather than throughout the entire effort. Some of the tools will not work at all for your company, and all can be carried to a self-defeating extreme (e.g., overelaboration of QFD, so that the house of quality becomes the mansion of quality). Again, you will have to determine what is right for your own situation and process. I have explicitly discussed the use of project management software and shown some examples because of its critical role in schedule and resource planning, both of which are central to shortening your time-to-profit.

Make use of appropriate tools at times when these can help you shorten time-to-profit.

In some cases you may find yourself forced to use a mandated tool or comply with a corporate policy in a way that may be unhelpful. You should certainly question the situation, but you may still have to go along.

In one company, a new product development effort often requires the design, construction, and installation of specialized production machinery, the cost of which must be capitalized. One project manager observed, "After the cost of a project has been estimated and the funds approved, there seems to be a requirement by management that the project be completed as close to that figure as possible—not running over or *under*" (project manager's emphasis). This has led to some cases in which equipment contractors have been asked to do work unrelated to that specific effort "just to use up excess funds." Clearly, in this case corporate policies have become straightjackets rather than helpful facilitation tools.

Avoid having policies that artificially constrain new product development projects.

SPECIAL DEVELOPMENT SITUATIONS

Several situations are encountered frequently and many practitioners feel that these are—or may be—substantially different than those for discrete manufactured goods, on which I have so far concentrated.

Radical Product Innovation

Radical product innovations arise from what has been defined [9] "as the propensity of a firm to introduce new products that: (1) incorporate substantially different technology from existing products; and (2) can fulfill key customer needs much better than existing products." Examples of radical innovations and what these replaced include word processors replacing typewriters, electronic calculators replacing slide rules, and cameras using a plastic-based film replacing those using a sensitive photographic emulsion on a glass plate. In such situations, prospective customers and users may not understand how a new technology can benefit them. The new product market opportunity may therefore be unclear. Consequently, the future payoff and its timing are uncertain, and responsible executives must have a willingness to tolerate considerable ambiguity and risk.

The willingness of a firm to cannibalize its own products with a radical innovation is inhibited by the extent of specialized investments in support of its existing products. Like the military, which is best able to fight the last war, many managers are proficient in yesterday's markets. This sets up its products for obsolescence and the firm for failure when there are market or technical discontinuities. Conversely, a firm is more likely to self-obsolete if there is paranoia about future markets, influential product champions for the radical innovation, or an internal market that allows initial testing [10].

Hewlett-Packard has avoided this reluctance to cannibalize its own products, as illustrated by the following statement: "We're obsoleting our own products by replacing them with better products before the competitors can" [11]. Similarly, Intel steadily introduces new microprocessor products that replace its existing chip products.

Radical product innovation causes markets to change rapidly. Table 5-1 illustrates some new product development practices that must be altered when working in such rapidly changing markets. Intel introduced a succession of its low-priced Celeron microprocessors in what appears to be a "make a little, sell a little" approach to find an attractive combination of features in this rapidly changing radical innovation market [12].

You must be alert if the markets in which your firm competes or the technologies upon which your products are based are now subject to rapid change or likely to become so. Be prepared to obsolete your own products. You should have FFE efforts under way as insurance at least.

Nonassembled Goods

Nonassembled products are homogeneous materials typically produced in bulk [13]. Examples of nonassembled goods include fluids that come from

TABLE 5-1. Summary of Effective Practices for New Product Development in Rapidly Changing Markets

Effective Practices in Rapidly Changing Markets	What's Different Compared to Development of Derivative Products in Stable Markets
Focus on probing, not measuring, in the NPD process. Involve users in idea generation, use mockups or prototypes to drive early and extensive qualitative market research targeted at understanding customer usage and benefits, target market segments, and the needs thereof.	Such probing may not be needed for derivative products in stable markets—user needs are well understood, market segments known.
Make a little, sell a little. Get early versions to market quickly, obtain user feedback, and modify as you go.	Such a practice would be risky in stable markets—competitors would "knock off" such products and beat the innovator into wider distribution with an improved product.
Limit large-scale, quantitative market research studies to answering questions of market size and best price points.	Market size and price points are typically well understood in these markets.
Stage the firm's resource commitments over the life of the NPD process. Involve top functional management early and often, and seek firm short-term and tentative long-term commitments of capital and human resources as the product concept evolves.	Risks are lower for derivative products. Top management support and involvement may be unnecessary to bring a derivative product to market.
Build an organizational culture that makes the development and introduction of successful new products everyone's job, to facilitate any necessary organizational and technological changes.	Organizational and technological changes are generally not necessary for derivative products.
Build direct marketing strategies for reaching desired markets which lessen the new product burden on the sales force and/or distribution channels.	Derivative products place fewer burdens on sales forces and channels, due to lesser needs to educate prospective users. Simple incentives can close the sale.

Source: *Journal of Product Innovation Management*, Volume 15 (May 1998), John W. Mullins and Daniel J. Sutherland, "New Product Development in Rapidly Changing Markets," pages 224–236, Copyright 1998, with permission from Elsevier Science.

a process stream (e.g., beverages, petroleum products, many chemicals, some medicines). Although these may subsequently be distributed in discrete packages, there are three important difference from discrete products: The production output is usually continuous for long periods of time; it is often shipped in railroad tank cars or ships, so plants are located close to appropriate transportation facilities; and the investment in production facilities may be capital intensive. But there really are tremendous similarities.

Rolls of paper, plastic sheeting, some metals, and most wire are discrete but produced on machinery that has a continuous output. As such, these are a hybrid with similarities to both discrete and nonassembled products. The requirement for continuous production means that the operations and production functions have tremendous importance. Trials and small lots of new products (or formulations) are intrusive to the income-producing output stream. It can be difficult to schedule such NPD efforts unless production downtime is deliberately included in operating budgets. In some cases (e.g., for adhesives products or where corrosive ingredients are required), cleanup of production machinery must be planned and may be time consuming. Safety and pollution control regulations must be considered in advance.

Software

New software products may be:

- A freestanding product on its own (e.g., mass market software for PCs)
- Embedded in a tangible product (e.g, the software to control an instrument or analyze the data collected by that instrument)
- Essential to support the operation of a new service (e.g., the back office software at a brokerage firm)

In many cases the design, development, debugging, and documentation of software is much slower than desired and this often becomes the critical path for time-to-market [14].

In the case of packaged software, developers seem to have a cultural addiction to lengthy periods of intense work and very long workdays in the weeks just before the scheduled launch. Software development in these conditions is more an artistic craft than a thoughtful, deliberate development aimed at systematically completing and debugging a user-friendly new product. Such a last-minute rush is undoubtedly a major contributor to the creation of errorware.

Three keys to speeding the development of software that will be embedded in hardware are:

1. Clear software requirements (which applies to all software, but especially this integrated class)
2. An integrated hardware–software development schedule
3. Testable software modules, delivered on time

The problem in many companies is exacerbated because many managers and executives lack understanding of or experience with software development, the software developers often have their own "free-spirited" culture, and today they commonly are a scarce resource.

Ideal support software is characterized by 3 Cs (compatibility, changeable, and crash resistant) and not having a fourth C (complexity). Sadly, nearly half of information technology software projects are stopped before they are implemented [15].

New Consumer Goods

Over 2000 new products are launched in the United States during a typical month, and the vast majority fail within a few years [16]. Some, such as Royal Crown's pioneering diet cola, fail because the market is not ready when it arrives. Only a very few create truly innovative and useful new product categories, such as Michelin's SA zero-pressure tire that will run some distance safely after a puncture or blowout. Many are aimed at the perilous fad market, which can reward a company for a fine product introduced at the right time and punish severely if it is not timely [17]. Point-of-purchase (POP) scanner data can be very important in tracking rates and patterns of purchase in consumer markets.

Procter & Gamble is trying to create a new category of cleanser because the traditional market is saturated and slow growing. They have identified fabrics that traditionally are dry cleaned as a target for their innovative fabric refresher, Febreze [18]. It is too early to determine whether or not this new product will be a major success.

The importance of advertising and product (and package) design is so great that it is common for advertising agencies and industrial design specialists to be part of the multifunctional NPD project team from the very beginning. Consumer market research is usually intensive and you should expect to invest in it significantly. This research should include both qualitative (i.e., focus groups, anthropological, etc.) and quantitative (i.e., conjoint analysis, survey data, etc.) forms. Large samples (hundreds or thousands) should be planned for rather than the handfuls (a dozen, 30, or sometimes 50) that are satisfactory for many business-to-business efforts. Finally, "hype" and fads can be very helpful and you should always look for ways to exploit these factors.

Services

People developing new services often feel that their situation is profoundly different from those who are developing tangible goods. Service examples include new banking products, new insurance products, communication services, education, health care, varied professional services, transportation, security protection, and so on. Specific examples include a certificate of deposit with unusual terms or guarantees, a policy to protect against the hazard of catastrophes in a risk-prone area, a new telephone rate plan, and so on.

Services have three important characteristics that distinguish them from tangible goods:

1. They are intangible and cannot be touched, so you may be unable to examine the service critically prior to use.
2. They are produced and consumed simultaneously and thus have no shelf life (except for database delivery services, such as credit reporting bureaus).
3. They are heterogeneous, produced and consumed at many locations (e.g., every hotel check-in desk), which means that quality can be highly variable.

In one case, a new service has been created by USAA Insurance Company that not only helps the user but may also save the provider considerable expense by reducing losses [19]. The concept is to have elderly drivers prepay for local-area taxicab rides so that they will not continue to drive their own cars. This service can save USAA Insurance losses because elderly drivers as a class are more prone to have accidents.

Unsatisfactory or disappointing services are—sadly—common. Because people like to exchange stories about the problems they encountered with a service (new or otherwise), these harmful messages proliferate far more rapidly than favorable reports about pleasant experiences. This means that the developer of a new service must invest heavily in testing the product and its delivery before its release. Using some employees as the test market frequently accomplishes this testing. It is also very important to invest in training the people who will deliver the service (e.g., the hotel or bank clerk).

SOURCES OF FURTHER HELP

No book can answer all questions about a topic. You may find yourself is a situation where you would like further personal help. Two organizations or members of these organizations may be useful to you.

Product Development & Management Association
Suite 100
236 Route 38 West
Moorestown, NJ 08057-3276
  TEL: (800) 232-5241
  FAX: (609) 231-4664
  Internet: http://www.pdma.org

New Product Development Specific Interest Group (NPD SIG)
Project Management Institute
Four Campus Boulevard
Newtown Square, PA 19073-3299
  TEL: (610) 355-4600
  FAX: (610) 355-4647
  Internet: http://www.pmi.org

SUMMARY OF KEY POINTS

- Resource overloading creates new product gridlock and should be avoided or reduced by establishing a clear NPD project priority system.
- Portfolio maps that array the characteristics of each project against its corresponding interval can help clarify if resources are being applied in a way that is consistent with strategy.
- Project management software can help identify potential resource overloads.
- Relationship maps can help clarify functional roles and responsibilities.
- Multifunctional teamwork is important and must be fostered.
- Special development situations generally follow the same principles but with additional or altered emphases.
- There are organizations and individuals that can provide your firm with assistance in implementing an improved NPD process to shorten your time-to-profit.

6

Continuous Improvement

INTRODUCTION

Whatever the details of your new product development process, the need for continuous improvement will soon be evident even if it is not yet painfully apparent. You will want—perhaps need—to extract the lessons learned from continuous improvement reviews and then disseminate these lessons to, and inculcate improvements in the organization and its new product development process. Such reviews are the most frequently omitted activity in new product development practices. Continuous improvement reviews require an investment of time immediately, but the payoff from this effort is both of uncertain value and will only be obtained sometime in the future. However, a key immediate benefit can be to avoid the repetition of avoidable mistakes or omissions.

> **Those who cannot remember the past are condemned to repeat it.**
> **George Santayana**

WHAT ARE CONTINUOUS IMPROVEMENT REVIEWS?

A continuous improvement review may have another label:

- Postmortem
- Project audit

- Postcommercialization review
- Postimplementation review
- Postcompletion review

These labels—and industry practice—are such that reviews are held *after* a new product development project ends, it they are conducted at all. The endpoint may be a market launch or some time after a market launch (to obtain the benefit of some customer and user reactions). More rarely, a review is held when a project ends because it is no longer sufficiently attractive, and this may truly be a postmortem.

It is probably best to adopt a neutral label to convey the purpose, which is continuous improvement. Thus, postmortem, with the connotation of death or a corpse that should be dissected might better be avoided, even though it is the most frequently observed label. An audit—regardless of the modifying adjective—has a connotation of financial review, which is only one aspect of a continuous improvement review. Postcommercialization review assumes that the product development effort achieved commercialization, but in many cases—the majority, in fact—this will not be the case because of the mortality of ideas and their concomitant projects. Postimplementation review is a common label for software development projects, but has the same problem as the postcommercialization review, in that many software development projects are terminated prior to implementation.

Postcompletion review or postproject appraisal are good labels, except that some development projects are inherently lengthy. Waiting until their completion to conduct a review to derive the lessons that may be learned precludes capitalizing on these lessons, which may be apparent much earlier and could be of immediate benefit to this or some other new product development project. In any event, because memories fade with time and some key people move on to other activities before a new product development project ends, continuous improvement reviews should be an ongoing activity throughout every new product development project [1]. So, although it may be more of a mouthful than some labels, continuous improvement review is probably the most descriptive label and therefore preferred.

Conduct continuous improvement reviews throughout the new product development effort, not just at the end.

WHY CONDUCT CONTINUOUS IMPROVEMENT REVIEWS?

The goal of a continuous improvement review is to learn from experience, to determine:

- What went right, so you can do it more frequently
- What went wrong, so you can modify new product development procedures or the plan for the project that you are now reviewing

In addition, you may be able to maximize some salvage value from a new product development project that must be suspended or terminated. A clear, complete, written record of previous continuous improvement reviews on a specific new product development project can help to jump start it if a resumption is justified.

Some of the ultimate benefits you can reasonably expect to obtain from the systematic conduct of continuous improvement reviews include:

- Better products
- More satisfied customers and users
- Lower product costs
- Shorter development schedules
- Reduced development expenses
- Generally improved business performance
- A more satisfied and enthusiastic staff of product developers

OBSTACLES TO CONTINUOUS IMPROVEMENT REVIEWS

Regrettably, there are many reasons that conspire against systematically conducting continuous improvement reviews. In fact, there are both tangible and intangible obstacles:

- The effort to conduct a continuous improvement review is required immediately, but the expected major benefits, if any, will occur primarily in the future.
- The people who must conduct the review are too busy with day-to-day activities or the next product development effort to devote sufficient attention to a review.
- There is lack of a widespread understanding about how to conduct a continuous improvement review.

- Schedules for key participants are incompatible.
- Time may be wasted.
- The review is perceived as a bureaucratic requirement rather than as a benefit for the participants.
- There is fear, justifiable or not, of delivering bad news to management.
- There is lack of motivation caused by the perceived expectations of senior management.
- There is potential for embarrassment (to self or friends).
- The possibility of confrontation, argument, or recrimination is feared.
- There is a tendency for finger-pointing, blame, and rebuke.
- Memory is perishable, and important data from earlier in the development effort are missing, misleading, or now believed to be incorrect.
- Key people with important insights are no longer available.

OVERCOMING THE OBSTACLES

Given these numerous and diverse obstacles, you must work proactively to establish a habitual practice of conducting continuous improvement reviews. Some ways that may help you to overcome these obstacles include:

- Institute continuous improvement reviews as an essential and integral part of the corporation's product development process.
- Initiate the practice of continuous improvement reviews with a new pilot product development project, modify the practice based on that experience, then apply it more widely.
- Establish—and maintain—a culture in which any "failure" is seen as a learning opportunity.
- Maintain a focus on the general lessons learned, and do not concentrate on or emphasize any person's mistakes.
- Provide praise and recognition for successful accomplishments.
- Use a skilled, neutral, objective person to manage the process and facilitate the majority—ideally all—of the continuous improvement reviews.
- Provide training to all staff about the objectives, benefits, procedures, and expected behaviors for continuous improvement reviews.
- Conduct continuous improvement reviews:
  - Just after market launch

- At appropriate times after the new product has been in use (e.g., 3, 6, and/or 12 months after market launch)
- At the end of product life
- Periodically throughout every new product development project
- More frequently

The kind of work being done during the NPD project calls for flexibility in the scheduling of reviews. During the FFE interval, there may be many small exploratory efforts that are deliberately chosen and authorized (e.g., a quick market research probe, a laboratory experiment or simple breadboard test). The results of these efforts cannot be predicted in advance. The next small FFE efforts to be undertaken frequently cannot be predicted or decided until the previous results are available. Thus, there may be many small projects, each of which deserves some formal review and documentation at its end. Conversely, the S&G interval is usually one comparatively large effort aimed at a clear objective (i.e., market launch). The end of this S&G interval is the most common time to conduct a review for companies that undertake these. After launch, during the PPS and CS intervals, there may be occasional efforts (hopefully small)—but often urgent—to correct some usage problem that has arisen unexpectedly. These are unintended efforts, especially if you have done the preceding work carefully and conscientiously. These are likely to be undertaken in a crisis "firefighting" mode because the firm is losing sales and profits while the precipitating problem remains unresolved. The resources used to deal with these unplanned PPS and CS efforts are likely to be taken from and delay another effort unless you have a function that is dedicated to the resolution of these usage problems. Each of these should conclude with a review.

The need for and timing of a review depend on the NPD interval.

Figure 6-1 illustrates some alternative timing options (arbitrarily showing four stages within the S&G interval, since it is most commonly divided into several stages). Option 1, the most frequently observed practice, is to conduct no reviews. Options 2 and 3 are to conduct a single postproject review, either at the time of launch or some time after launch when early field experience is available. For those who do conduct continuous improvement reviews, option 2 or 3 is the most common pattern. Option 4 is to conduct a review at the end of each interval (FFE, S&G, PPS, and CS) as well as the end of each major stage or milestone within the S&G interval. Because the time between reviews in option 4 can be lengthy, adding *periodic* (e.g., monthly)

Conduct both episodic and periodic continuous improvement reviews.

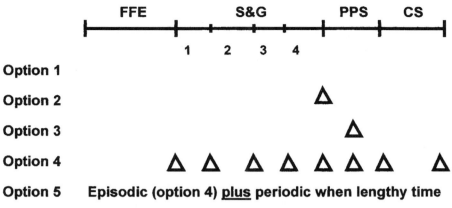

Figure 6-1. Options for timing of continuous improvement reviews.

reviews between the *episodic* ones illustrated in option 4 is prudent—and option 5 is recommended.

KINDS OF INFORMATION TO OBTAIN

Three actions are involved in useful continuous improvement reviews:

1. Collecting pertinent information
2. Deriving learnings from that information
3. Disseminating the learnings

There is an obvious challenge in deciding what information will be pertinent before it is collected. The goal—seldom realized in practice—is to collect minimum information that will provide maximum insight. However, in most companies striving for continuous improvement, more information is collected than is actually useful. Two kinds of information must be examined—information about individual projects and the entire portfolio of projects—and both are discussed in the next two sections.

Information About an Individual Project

It is almost always helpful to collect frequent (e.g., monthly) measures of whatever changes occur in key aspects of the:

· Expected product performance

- Anticipated schedule for time-to-market and time-to-profit
- Actual versus planned development effort.

Whenever a change in any of these three aspects is detected, causative information should also be sought. The objective is to determine whether a change is due to people (e.g., organization, availability, skill mix) or to things (e.g., procedures, business practices, equipment and facility availability, suppliers). Constantly try to determine what could have been done better if it had been done differently.

Information About the Entire Portfolio of Projects

The basic objective in this category of data collection and examination is to determine first if the portfolio properly supports the overall corporate strategy. You will want to determine if the new product development project mix is consistent with the balance of risk, appropriately dedicated to promising business areas, showing a suitable distribution of probable times for market entry, and capitalizing on a reasonable—ideally, optimum—use of corporate resources.

Second, you will want to examine collective data about the progress of all projects through the new product development process. This might include such measures as:

- The number of projects entering each interval (FFE, S&G, PPS, and CS) and if there are multiple stages within the S&G interval, each stage within the S&G interval
- The number of projects exiting each stage and how long they were in the stage
- Some measure of the effort (full-time equivalent labor, labor hours, money, or some proxy for these items) in each stage

The objective of these measures is to try to identify any persistently "sticky" intervals or stages so that the overall process can be improved.

The third category of collective data is the overall performance of the portfolio of new product development efforts. This might be improvement in the percentage of orders, sales, or profits attributed to new products introduced in the last two to five years. Exemplars include 3M, Johnson & Johnson, and Hewlett-Packard, as cited in Chapter 1. For this measure to be meaningful, the definition of what constitutes a new product must be unambiguous and stable for many years; otherwise, managers will downgrade what it requires to be "new" and thus improve the track record.

Some people [2] advocate looking at the improvement in the cumulative new product income from products introduced in the last few years compared to the cumulative investment in new product development in that same period. The problem with this class of measures is the inherent time lag between investment, and income can be very variable even within a single company or strategic business unit (SBU). Metrics must be meaningful, a snapshot in time that measures how well the company is doing and the health of its processes, with trends over time clearly revealed. Useful metrics are valid and reliable, defined unambiguously, ideally dependent on data that are economic to collect, available in a timely way—and acted upon promptly.

Collect information about individual projects and the entire portfolio of projects.

HOW TO CONDUCT CONTINUOUS IMPROVEMENT REVIEWS

A successful program of benefiting from continuous improvement reviews can start and be maintained only with the active and wholehearted support of senior executives. They must support both the idea and the effort, and they must establish and maintain a culture in which any "failure" is recognized and understood to be a learning opportunity. Once these basic conditions are met, there are a myriad of more mechanical elements that must be considered:

- The facilitator and any auditor(s) must be respected, mature, neutral, and objective. This is critical so those findings are not rejected because the facilitator or an auditor has a marginal reputation. Whether the facilitator and auditors should be hired from outside the company (which promotes objectivity) or from inside (which may improve familiarity with the company's procedures, business, products, and technology) depends on the situation. Other selection criteria for the facilitator and auditors include:
  - Designation by a senior executive
  - Acceptable to project team
  - Pertinent experience
  - Ability to ask questions, especially nondirective ones
  - Ability to write reports, communicate, and present results
- Objectivity is required to assure that the result is learning, not blame.
- Periodic continuous improvement reviews are highly desirable during the course of any new product development project and are an absolute requirement throughout a lengthy effort because:

- Definitive findings can be acted on immediately.
- Other information can be stored for use in subsequent continuous improvement reviews as more history is gathered.
- Personnel who may have to actively participate vary depending on the specific situation, but may include:
  - Members of the multifunctional project team
  - Functional managers, process owners, executives
  - Critical suppliers
  - Key lead customers and users
- One-to-one interviews with key personnel must be conducted in a neutral, nonthreatening environment. A skilled facilitator will have a prepared list of initial questions that can be and are modified as different people are queried for their input. The option for reinterviewing must be available to clarify points in light of new information collected from others.

Finally, the results of every continuous improvement review must be disseminated, and new product development practices have to be either validated or changed. That is, the findings must be implemented. A final report that is stored and not acted upon is a waste of scarce resources. Two effective ways to communicate identified improvement possibilities rapidly are by use of the company's electronic communication network or through process champions. In some cases, training sessions, perhaps including the participation of key suppliers or even lead users, will be appropriate to inculcate new learnings or explain new procedures.

> **The findings from continuous improvement reviews must be believed and acted upon.**

There are some other ways to transfer knowledge. The team that is about to start a NPD project might audit several current projects that are at various stages of progress. You may find it practical to staff your next NPD project with some team members who have recently completed work on another NPD project and have these experienced personnel conduct a startup workshop for the less experienced team members. Or, you may hire one or a few people who have learned by failure on other efforts, since they have been educated and will be smarter the next time.

Hewlett-Packard's corporate engineering function measures many aspects of NPD (and other) work and reports the results widely throughout the corporation.

The Product Development & Management Association (PDMA) holds international conferences annually, organized and largely managed by a team of volunteers who change from one year to the next. With only a few exceptions, where the chairperson serves a second time (but several years later), the volunteers are different each time. These PDMA conferences are major undertakings and the transference of knowledge from one set of volunteers to the next was idiosyncratic in that some chairpersons compiled extensive written reports, and others relied primarily on a verbal briefing of his or her successor. Although there have been general constants as to the conference, accompanying workshops, and other events, each conference is held in a new city, a new hotel, run by a new team, and includes new and untested program elements. To improve the transfer of knowledge, the chairperson of the conference to be held two years in the future chairs the continuous improvement review conducted at the end of the current year's conference with the current and next-year's teams.

SUMMARY OF KEY POINTS

- Continuous improvement reviews have frequently been inappropriately called postmortems
- The two goals of continuous improvement reviews are to capitalize on new and better ways to carry out NPD that have been devised and to avoid repeating ineffective practices.
- There are many obstacles to these important reviews, the most prevalent being that the people involved are too busy on other NPD projects.
- To overcome the obstacles, continuous improvement reviews must be seen as an integral part of the company's NPD process.
- Information has to be collected and disseminated about both individual projects and the portfolio of NPD projects.
- Facilitators can be helpful in conducting the reviews.

Appendices

Appendix A: Abbreviations

ABC	Activity-based costing
CAD	Computer-aided design
CAE	Computer-aided engineering
CAM	Computer-aided manufacturing
CASE	Computer-aided software engineering
CIR	Continuous improvement review
CS	Continuing sales (interval)
D&D	Design and development
DCF	Discounted cash flow
DFMA	Design for manufacturing and assembly
DOE	Design of experiments
DRAM	Dynamic random-access memory
FEMA	Failure mode and effects analysis
FFE	Fuzzy front end (interval)
IRR	Internal rate of return
JIT	Just-in-time
NHTSA	National Highway Traffic Safety Administration
NIH	Not invented here
NPD	New product development
NPV	Net present value
PC	Personal computer
PM	Project manager
POP	Point of purchase
POS	Point of sales
PPS	Preprofit sales (interval)

PR	Public relations
QFD	Quality function deployment
R&D	Research and development
RD&E	Research, development, and engineering
S&G	Stages and gates (interval)
SBU	Strategic business unit
TCD&D	Time-critical design and development
3-D	Three-dimensional
TQM	Total quality management
WOM	Word of mouth

Appendix B: References

Chapter 1

[1] S. Manes, "Have Bugs, Will Deliver (Help! Help!)," *The New York Times,* June 9, 1998, page B14.

[2] J. Kalov, "Frustrating the Customer," *Chief Executive,* June 1997, pages 52–53.

[3] *3M 1997 Annual Report,* page 4.

[4] *Johnson & Johnson 1997 Annual Report,* inside front cover and page 1.

[5] See, for example, *1995 and 1996 Hewlett-Packard Annual Reports.*

[6] L. Napoli, "Mourning the End of a Printing Era," *The New York Times,* March 20, 1998, page YC5.

[7] F. Popoff, "What's the Secret of Success?" presentation at the Spring Meeting of the Commercial Development Association, Chicago, March 3–6, 1996.

[8] Numerous examples have been cited. It is always uncertain that previous development time has been measured in a way that is strictly similar to the improvement cited. Nevertheless, some time reductions are so great that precise measurement comparability is relatively unimportant. See, for example, A. Barrett and G. DeGeorge, "Home Improvement at Black & Decker," *Business Week,* May 11, 1998, page 54ff, which cites a reduction of new product development time from 36 months to 20 months over a five-year period.

[9] Anon., "Nissan Plan to Speed Auto Development," *The New York Times,* December 27, 1997, page Y3B.

[10] R. G. Cooper, *Winning at New Products,* second edition. Reading, MA: Addison-Wesley, 1993.

[11] J. Arleth, "Using PROBE to Benchmark Your New Product Process," *ISPIM Newsletter,* April 1998, page 14.

[12] C. Goldsmith, "Boeing Plans to Make 'Drastic' Changes to Slash Development Time for Aircraft," *The Wall Street Journal,* February 11, 1998, page B2.

[13] B. L. Bayus, "Speed-to-Market and New Product Performance Trade-offs," *Journal of Product Innovation Management,* November 1997, pages 485–497.

[14] A. Griffin and A. L. Page, "PDMA Success Measurement Project: Recommended Measures for Product Development Success and Failure," *Journal of Product Innovation Management,* November 1996, pages 478–496.

[15] R. Minter, "The Myth of Market Share," *The Wall Street Journal,* June 15, 1998, page A28.

[16] C. H. House and R. L. Price, "The Return Map," *Harvard Business Review,* January– February 1991, pages 92–100.

Chapter 2

[1] This model was developed at a meeting of practitioners, academics, and service providers in Lakeville, CT, in May 1997.

[2] Anon., "Here, Innovation Is No Fluke," *Fast Company,* August–September 1997, page 42ff.

[3] D. Takahashi, "Doing Fieldwork in the High-Tech Jungle," *The Wall Street Journal,* October 27, 1998, page B1ff.

[4] E. Davenport, "Building on the Eureka Factor: Taking New Products to Market," presentation at the Spring Meeting of the Commercial Development Association, Chicago, March 3, 1996.

[5] M. Maremont, "New Toothbrush Is Big-Ticket Item," *The Wall Street Journal,* October 27, 1998, page B1.

[6] O. Port, "Warning: Red Means Frozen," *Business Week,* July 6, 1998, page 85.

[7] A. Latour, "Sheet Speakers May Be on the Wall," *The Wall Street Journal Europe,* November 13–14, 1998, page 4.

[8] L. Gomes, "It Sounded So Good: The History of Consumer Electronics Is Littered with Failure," *The Wall Street Journal,* June 11, 1998, page R22; A. Griffin, "Obtaining Customer Needs for Product Development," Chapter 11 in M. D. Rosenau et al., editors, *The PDMA Handbook of New Product Development* (New York: Wiley, 1996), pages 153–168.

[9] S. Warren, "Whatever Happened to the Buckyball?" *The Wall Street Journal,* May 4, 1998, page B1ff.

[10] M. Ritter, "Electronic Ink Shifts Shapes Like Magic," *Houston Chronicle,* December 4, 1998, page 2D.

[11] M. Kenward, "Displaying a Winning Glow," *Technology Review,* January–February 1999, page 68ff.

[12] J. Markoff, "F.C.C Mulls Wider Commercial Use of Radio," *The New York Times,* December 21, 1998, page C3ff.

[13] D. Takahashi, "Intel to Offer Computer-Help Service to PC Users," *The Wall Street Journal,* March 11, 1998, page B5; P. H. Lewis, "Only the Answers Are Free," *The New York Times,* March 19, 1998, page YD3; Anon, "Intel Launches Pay Service for Computer Help," *Houston Chronicle,* March 11, 1998, page 3C.

[14] L. Bransten, "E-Stamp Aims to Lick Postal Tradition," *The Wall Street Journal,* April

16, 1998, page B10; J. G. Auerbach, "Meter Maker's Suits Claim Postage Due," *The Wall Street Journal,* July 6, 1998, pages A17ff; G. Anders, "It's Digital, It's Encrypted—It's Postage," *The Wall Street Journal,* September 21, 1998, page B1.

[15] F. J. Prial, "No More Free Drinks: System Keeps Tabs on Bartenders," *The New York Times,* March 19, 1998, page D10Y.

[16] K. Santos, "Compaq Unveils Fingerprint ID for Computer Users," *Houston Chronicle,* July 8, 1998, page 1Cff.

[17] S. C. Wheelwright and W. E. Sasser, Jr., "The New Product Development Map," *Harvard Business Review,* May–June 1989, pages 112–125.

[18] V. H. Prushan, "Better Off Dead?" *Entrepreneur,* April 1998, pages 128–130.

[19] D. J. Shelley et al., "A Lower-Cost Inkjet Printer Based on a New Printing Performance Architecture," *Hewlett-Packard Journal,* June 1997, pages 6–11.

[20] See *Hewlett-Packard Company 1995 Annual Report,* page 13.

[21] W. S. Mossberg, "The CrossPad Sends Paper-and-Ink Notes to Your PC Screen," *The Wall Street Journal,* April 9, 1998, page B1; S. H. Wildstrom, "A Clipboard with a Memory," *Business Week,* April 27, 1998, page 20.

[22] See, for example, two fine books by E. F. McQuarrie, *The Market Research Toolbox: A Concise Guide for Beginners* (Thousand Oaks, CA: Sage Publications, 1996), and *Customer Visits: Building a Better Market Focus,* second edition (Thousand Oaks, CA: Sage Publications, 1998). Focus groups are well covered in T. L. Greenbaum, *The Handbook of Focus Group Research,* second edition (Thousand Oaks, CA: Sage Publications, 1998). A basic explanation of conjoint analysis is provided in M. D. Rosenau, *Faster New Product Development* (New York: Amacom, 1990), Appendix B, pages 239–257.

[23] G. H. Loosschilder, M. J. W. Stokmans, and D. R. Wittnik, "Convergent Validity of Preference Estimates of the Interactive Concept Test," paper presented at the 21st Annual International Conference of the Product Development & Management Association, Monterey, CA, October 1997.

[24] L. P. Feldman, "The Perils of Grabbing Lightning," *Visions,* April 1998, page 11ff. See also, G. C. O'Connor, "Market Learning and Radical Innovation: A Cross Case Comparison of Eight Radical Innovation Projects," *Journal of Product Innovation Management,* March 1998, pages 151–166.

[25] X. M. Song and M. M. Montoya-Weiss, "Critical Development Activities for Really New Versus Incremental Products," *Journal of Product Innovation Management,* March 1998, pages 124–135.

[26] R. W. Veryzer, Jr., "Key Factors Affecting Customer Evaluation of Discontinuous New Products," *Journal of Product Innovation Management,* March 1998, pages 136–150.

[27] I am grateful to A. J. Dankworth, who suggested this example in a talk, "Business Opportunity Analysis and the Project Team," at a meeting of the Commercial Development Association, Houston, September 29, 1997.

[28] V. K. Jolly, *Commercializing New Technologies: Getting from Mind to Market.* Boston: Harvard Business School Press, 1998, page 365.

[29] G. McCool, "Sun-Fueled Electric Cars Hit the Road," *Houston Chronicle,* May 11, 1998, page 7A; N. Antosh, "Prospects Dimmer for Electric Vehicles," *Houston Chronicle,* December 8, 1998, page 3C; A. Pollock, "Charge! Doing an Electric Commute," *The New York Times,* July 26, 1998, page 30Y.

[30] M. Krebs, "Planting the Seeds for a Crop of Lean, Green Machines," *The New York Times,* March 15, 1998, page Y39.

[31] *Eli Lilly & Company 1997 Annual Report,* pages 3 and 12.

[32] *C.R. Bard, Inc. 1997 Annual Report,* pages 4–5.

[33] *Merck & Company 1997 Annual Report,* page 2.

[34] G. Haman, "Techniques and Tools to Generate Breakthrough New Product Ideas." Chapter 12 in M. D. Rosenau et al, editors, *The PDMA Handbook of New Product Development* (New York: Wiley, 1996), pages 167–178. See also, P. C. Judge, "Artificial Imagination," *Business Week,* March 18, 1995, page 60; T. Proctor, "New Development Is Computer Assisted Creative Problem Solving," *Creativity and Innovation Management,* June 1997, pages 94–98.

[35] P. L. Zweig and W. Zellner, "Locked Out of the Hospital," *Business Week,* March 16, 1998, pages 75–76; see also, Letter to the Editor from C. Castellini, "Becton Dickinson: Safe Needles Are Easy to Find," *Business Week,* April 6, 1998, page 11.

[36] S. D. Moore, "Failure Is No Stranger in Drug Research," *The Wall Street Journal,* June 15, 1998, page B10D.

[37] G. A. Stevens and J. Burley, "3,000 Raw Ideas = 1 Commercial Success!" *Research-Technology Management,* May–June 1997, pages 16–27.

[38] R. J. Thomas, "New Product Success Stories: Lessons from Leading Innovators," presentation at the Spring Meeting of the Commercial Development Association, Chicago, March 3–6, 1996.

[39] A. Griffin et al., *Drivers of NPD Success: The 1997 PDMA Report.* Chicago: Product Development & Management Association, 1997, page 5.

[40] F. Rose, "Electronic-Nose Concern Seeks Sweet Smell of Success," *The Wall Street Journal,* April 30, 1998, page B8.

[41] See, for example, R. S. Kaplan and R. Cooper, *Cost & Effect: Using Integrated Cost Systems to Drive Profitability and Performance* (Boston: Harvard Business School Press, 1997); D. Berlant, R. Browning, and G. Foster, "How Hewlett-Packard Gets Numbers It Can Trust," *Harvard Business Review,* January–February 1990, page 178ff.

Chapter 3

[1] See, for example; J. R. Hauser and D. Clausing, "The House of Quality," *Harvard Business Review,* May–June 1988, pages 63–73; or M. D. Rosenau and J. J. Moran, *Managing the Development of New Products* (New York: Van Nostrand Reinhold, 1993), pages 225–237.

[2] S. H. Wildstrom, "This Cool Laptop Is Hard to Use," *Business Week,* June 15, 1998, page 26.

[3] This relationship is suggested in W. E. Souder, J. D. Sherman, and R. Davies-Cooper, "Environmental Uncertainty, Organizational Integration, and New Product Development Effectiveness: A Test of Contingency Theory," *Journal of Product Innovation Management,* November 1998, pages 520–533. Some related issues are covered in C. Karlsson, R. Nellore, and K. Soderquist, "Black Box Engineering: Redefining the Role of Product Specifications," *Journal of Product Innovation Management,* November 1998, pages 534–549.

[4] C. H. Deutsch, "Six Sigma Enlightenment: Managers Seek Corporate Nirvana Through Quality Control," *The New York Times,* December 7, 1998, page C1Yff.

[5] D. Lester, "Problems with New Product Development," paper presented at the 20th Annual International Conference of the Product Development & Management Association, Orlando, FL, October 1996.

[6] M. D. Rosenau, "Choosing a Development Process That's Right for Your Company," Chapter 6 in M. D. Rosenau et al., editors, *The PDMA Handbook of New Product Development.* New York: Wiley, 1996, page 85.

[7] M. D. Rosenau, *Successful Project Management,* third edition. New York: Wiley, 1998.

[8] Rosenau, "Choosing a Development Process That's Right for Your Company," pages 88 and 89.

Chapter 4

[1] S. H. Wildstrom, "Why Microsoft Must Do Better," *Business Week,* March 30, 1998, page 20.

[2] R. O. Crockett, "Wireless Goes Haywire at Motorola," *Business Week,* March 9, 1998, page 32.

[3] See *AAdvantage Newsletter,* March–April 1998, page 4.

[4] M. Krebs, "Is Cadillac Prepared to Take On the World?" *The New York Times,* February 15, 1998, page 38Y.

[5] N. Gross, "High Anxiety from High-Definition TV," *Business Week,* March 23, 1998, page 89.

[6] Anon., "New Web-site Finder Simply Isn't Working as Intended," *Houston Chronicle,* March 13, 1998, page 5C.

[7] Anon., "Court Rejects Chrysler's Recall Appeal," *Houston Chronicle,* March 21, 1998, page 3C.

[8] K. Goffin, "Evaluating Customer Support During New Product Development—An Empirical Study," *Journal of Product Innovation Management,* January 1998, pages 42–57.

[9] See, for example, C. Roberts, "Tech Support Puts Help on the Line," *Houston Chronicle,* February 8,1998, page 5Eff.; S. Manes, "Settlement Near in Technical Help-line Suit," *The New York Times,* March 3, 1998, page B12Y; and Anon., "Do Computers Have to Be Hard to Use?" *The New York Times,* May 28, 1998, page YD1ff.

[10] L. Johannes, "For New Film, A Brighter Picture," *The Wall Street Journal,* May 5, 1998, page B1ff. I have also benefited from discussions at a meeting in Interlaken, CT, in May 1998 with practitioners involved in this development.

[11] M. D. Rosenau and J. J. Moran, *Managing the Development of New Products,* New York: Van Nostrand Reinhold, 1993.

[12] T. Petzinger, Jr., "It Takes Humility to Market Even the Hottest Products," *The Wall Street Journal,* March 27, 1998, page B4.

[13] T. L. Burton, "Baxter Suspends Trial of Blood Substitute," *The Wall Street Journal,* June 3, 1998, page B11.

[14] E. J. Hultnik, A. Griffin, S. Hart, and H. S. J. Robben, "Industrial New Product Launch Strategies and Product Development Performance," *Journal of Product Innovation Management,* July 1997, pages 243–257.

[15]   M. L. Fisher, "What Is the Right Supply Chain for Your Product?" *Harvard Business Review,* March–April 1997, pages 104–117.

[16]   L. Johannes, "Kodak Looks for Another Moment with Film for Pros," *The Wall Street Journal,* March 13, 1998, page B1.

[17]   K. Atuahene-Gima and H. Li, "The Determinants of Salespeople's Commitment to Sell New Products: An Empirical Analysis," paper presented at 21st Annual International Conference of the Product Development & Management Association, Monterey, CA, October 1997.

[18]   N. Rackham, "From Experience: Why Bad Things Happen to Good New Products," *Journal of Product Innovation Management,* May 1998, pages 201–207.

[19]   C. M. Crawford, *New Products Management,* Burr Ridge, IL: Irwin, 1994.

[20]   D. Canady, "Gillette Unveils Its Mach 3 Razor as Stock Backs Off," *The New York Times,* April 15, 1998, page YC1ff.; M. Maremont, "How Gillette Brought Its Mach 3 to Market," *The Wall Street Journal,* April 15, 1998, page B1ff.; W. C. Symonds, "Would You Spend $1.50 for a Razor Blade?" *Business Week,* April 27, 1998, page 46; J. Surowiecki, "The Billion-Dollar Blade," *The New Yorker,* June 15, 1998, pages 43–49; G. G. Marcial, "An Even Sharper Edge on Gillette," *Business Week,* July 20, 1998, page 116; L. Johannes, "Gillette Co. Sees Strong Early Sales for Its New Razor," *The Wall Street Journal,* July 17, 1998, page B3.

[21]   C. H. Deutsch, "Keeping Computer Printers in Ink," *The New York Times,* April 20, 1998, page YC5.

[22]   D. Canady, "Where Nothing Lasts Forever," *The New York Times,* April 24, 1998, page YC1ff.

[23]   C. H. Deutsch, "A Road Warrior of Innovators," *The New York Times,* March 28, 1998, page B1Yff.

[24]   M. Inasiti, *Technology Integration.* Boston: Harvard Business School Press, 1998, page 51.

[25]   V. K. Jolly, *Commercializing New Technologies.* Boston: Harvard Business School Press, 1997, pages 287–288.

[26]   W. S. Mossberg, "Personal Technology," *The Wall Street Journal,* February 12, 1998, page B1.

[27]   M. Stepanek, "What Smart Cards Couldn't Figure Out," *Business Week,* November 30, 1998, page 142.

[28]   Anon., "Kodak to Stop Making Film for Its Instamatics," *The Wall Street Journal,* May 7, 1998, page A11.

[29]   P. A. Katzfey, "Product Discontinuation," Chapter 28 in M. D. Rosenau et al., editors, *The PDMA Handbook of New Product Development.* New York: Wiley, 1996, pages 413–424.

[30]   J. Griffin, *Customer Loyalty.* San Francisco: Jossey-Bass, 1997.

Chapter 5

[1]   J. J. McKeown, "New Products from New Technologies," *Journal of Business & Industrial Marketing,* Volume 5, Number 1, Winter–Spring 1990, pages 67–72.

[2]   *3M Company 1997 Annual Report,* page 5.

[3]   M. D. Rosenau, *Successful Project Management,* third edition. New York: Wiley, 1998, especially Chapter 25, pages 307–318.

[4]   G. A. Rumler and A. P. Brache, *Improving Performance,* second edition. San Francisco: Jossey-Bass, 1995, especially pages 38 and 189. Also, G. S. Tighe, "Using Concurrent Teams to Re-engineer the Product Development Process" in *Unraveling the Complexity of New Product Design, Development and Commercialization.* Indianapolis, IN: Product Development & Management Association, 1995, pages 11–17.

[5]   R. G. Cooper, S. J. Edgett, and E. J. Kleinschmidt, *"Portfolio Management for New Products,"* Hamilton, Ontario: McMaster University, 1997.

[6]   Rosenau, *Successful Project Management,* pages 200–205.

[7]   O. Suris, "Behind the Wheel," *The Wall Street Journal,* November 18, 1996, page R14ff.

[8]   J. G. Kunkel, "Rewarding Product Development Success," *Research-Technology Management,* September–October 1997, pages 29–31.

[9]   R. K. Chandy and G. J. Tellis, "Organizing for Radical Product Innovation: The Overlooked Role of Willingness to Cannibalize," *Journal of Marketing Research,* 35 (November 1998), pages 474–487.

[10]   R. K. Chandy, "Radical Product Innovation," presentation at the Product Development & Management Association South-Central Chapter Meeting, November 11, 1997.

[11]   *Hewlett-Packard 1995 Annual Report,* page 13.

[12]   L. M. Fisher, "Pentium Chip Prices Sliced 12% to 20%," *The New York Times,* June 9, 1998, page C6Y.

[13]   T. D. Parrish and L. E. Moore, "Nonassembled Product Development," Chapter 20 in M. D. Rosenau et al., editors, *The PDMA Handbook of New Product Development.* New York: Wiley, 1996, pages 287–295.

[14]   E. Carmel, "Cycle Time in Packaged Software Firms," *Journal of Product Innovation Management,* Volume 12, Number 2 (March 1995), pages 110–123; T. G. Rauscher and P. G. Smith, "From Experience: Time-Driven Development of Software in Manufactured Goods," *Journal of Product Innovation Management,* Volume 12, Number 3 (June 1995), pages 185–199.

[15]   B. Wysocki, "Pulling the Plug: Some Firms, Let Down by Costly Computers, Opt to 'De-engineer,' " *The Wall Street Journal,* April 30, 1998, page 1ff.

[16]   B. Kanner, "Marketers Pick Likely Winners Out of New Products," *Houston Chronicle,* March 22, 1998, page 2D.

[17]   D. P. Hamilton, "Sega to Take Write-off and Post Loss, Illustrating Its Missteps in Video Games," *The Wall Street Journal,* March 16, 1998, page A18.

[18]   T. Parker-Pope, "P&G Targets Textiles Tide Can't Clean," *The Wall Street Journal,* April 29, 1998, page B1ff.

[19]   C. Terhune, "USAA Tries Taxi Plan for Elderly," *The Wall Street Journal,* May 13, 1998, page T1ff.

Chapter 6

[1]   The topic is not well treated in the literature, but some coverage exists: M. D. Rosenau, *Successful Project Management: A Step-by-Step Approach with Practical Examples,* third edition (New York: Wiley, 1998), pages 295–296 and 312–313; P. G. Smith and D. G. Reinertsen, *Developing Products in Half the Time: New Rules, New Tools* (New

York: Wiley, 1998), pages 286–289; S. C. Wheelwright and K. B. Clark, *Revolutionizing Product Development: Quantum Leaps in Speed, Efficiency, and Quality* (New York: Free Press, 1992), pages 284–310.

[2]   See, for example, T. D. Kuczmarski, *Innovation: Leadership Strategies for the Competitive Edge* (Lincolnwood, IL: NTC Business Books, 1996), page 185; or M. E. McGrath and M. N. Romeri, "The R&D Effectiveness Index: A Metric for Product Development Performance," *Journal of Product Innovation Management,* June 1994, pages 213–220.

Appendix C: Recent Helpful Books on Aspects of New Product Development

Bacon, Frank R., Jr., and Butler, Thomas W., Jr.
Achieving Planned Innovation: A Proven System for Creating Successful New Products and Services
New York: Free Press, 1998

Brown, Shona L., and Eisenhardt, Kathleen M.
Competing on the Edge: Strategy as Structured Chaos
Boston: Harvard Business School Press, 1998

Christensen, Clayton M.
The Innovator's Dilemma: When New Technologies Cause Great Firms to Fail
Boston: Harvard Business School Press, 1997

Cooper, Robert G., Edgett, Scott J., and Kleinschmidt, Elko J.
Portfolio Management for New Products
Reading, MA: Addison-Wesley, 1998

Dimancescu, Dan, and Dwenger, Kemp
World-Class New Product Development: Benchmarking Best Practices of Agile Manufacturers
New York: Amacom, 1996

Graham, Robert J., and Englund, Randall L.
Creating an Environment for Successful Projects: The Quest to Manage Project Management
San Francisco: Jossey-Bass, 1997

Jolly, Vijay K.
Commercializing New Technologies: Getting from Mind to Market
Boston: Harvard Business School Press, 1997

Kuczmarski, Thomas D.
Innovation: Leadership Strategies for the Competitive Edge
Lincolnwood, IL: NTC Business Books, 1995

McGrath, Michael E., editor
Setting the PACE in Product Development: A Guide to Product and Cycle-Time Excellence
Boston: Butterworth-Heinemann, 1996

McQuarrie, Edward F.
The Market Research Toolbox: A Concise Guide for Beginners
Thousand Oaks, CA: Sage Publications, 1996

McQuarrie, Edward F.
Customer Visits: Building a Better Market Focus, second edition
Thousand Oaks, CA: Sage Publications, 1998

Meyer, Mark H., and Lehnerd, Alvin
The Power of Product Platforms: Creating and Sustaining Robust Corporations
New York: Free Press, 1997

Pisano, Gary P.
The Development Factory: Unlocking the Potential of Process Innovation
Boston: Harvard Business School Press, 1997

Reinertsen, Donald
Managing the Design Factory: A Product Developer's Toolkit
New York: Free Press, 1997

Rosenau, Milton D., Jr.
Successful Project Management: A Step-by-Step Approach with Practical Examples (third edition)
New York: Wiley, 1998

Rosenau, Milton D., Jr. et al., editors
The PDMA Handbook of New Product Development
New York: Wiley, 1996

Smith, Preston G., and Reinertsen, Donald G.
Developing Products in Half the Time: New Rules, New Tools
New York: Wiley, 1998

Thomas, Robert J.
New Product Success Stories: Lessons from Leading Innovators
New York: Wiley, 1995

Tushman, Michael L., and O'Reilly, Charles A., III
Winning Through Innovation: A Practical Guide to Leading Organizational Change
Boston: Harvard Business School Press, 1996

Index